Effective Science Communication (Second Edition)

A practical guide to surviving as a scientist

Effective Science Communication (Second Edition)

A practical guide to surviving as a scientist

Sam Illingworth
School of Biological Sciences, The University of Western Australia, Perth, Australia

Grant Allen
School of Earth and Environmental Science, University of Manchester, Manchester, UK

IOP Publishing, Bristol, UK

© IOP Publishing Ltd 2020

All rights reserved. No part of this publication may be reproduced, stored in a retrieval system or transmitted in any form or by any means, electronic, mechanical, photocopying, recording or otherwise, without the prior permission of the publisher, or as expressly permitted by law or under terms agreed with the appropriate rights organization. Multiple copying is permitted in accordance with the terms of licences issued by the Copyright Licensing Agency, the Copyright Clearance Centre and other reproduction rights organizations.

Permission to make use of IOP Publishing content other than as set out above may be sought at permissions@ioppublishing.org.

Sam Illingworth and Grant Allen have asserted their right to be identified as the authors of this work in accordance with sections 77 and 78 of the Copyright, Designs and Patents Act 1988.

ISBN 978-0-7503-2520-2 (ebook)
ISBN 978-0-7503-2518-9 (print)
ISBN 978-0-7503-2521-9 (myPrint)
ISBN 978-0-7503-2519-6 (mobi)

DOI 10.1088/978-0-7503-2520-2

Version: 20200501

IOP ebooks

British Library Cataloguing-in-Publication Data: A catalogue record for this book is available from the British Library.

Published by IOP Publishing, wholly owned by The Institute of Physics, London

IOP Publishing, Temple Circus, Temple Way, Bristol, BS1 6HG, UK

US Office: IOP Publishing, Inc., 190 North Independence Mall West, Suite 601, Philadelphia, PA 19106, USA

For Becky and Ian, whose love and support enable us to do the jobs that we love.

Contents

Second edition preface	xi
Acknowledgements	xii
Author biographies	xiii

1 Introduction — 1-1
1.1 Introduction — 1-1
1.2 How to use this book — 1-2
1.3 Summary — 1-4
1.4 Further study — 1-4
1.5 Suggested reading — 1-5
    References — 1-5

2 Publishing work in academic journals — 2-1
2.1 Introduction — 2-1
2.2 Scoping your deliverables — 2-2
2.3 Choosing a journal — 2-5
2.4 Writing and manuscript preparation — 2-7
2.5 The peer review process — 2-9
2.6 Reviewing papers — 2-12
2.7 Citations and metrics — 2-14
2.8 Summary — 2-16
2.9 Further study — 2-16
2.10 Suggested reading — 2-17
    References — 2-17

3 Applying for funding — 3-1
3.1 Introduction — 3-1
3.2 What makes a good idea? — 3-2
3.3 Finding funding — 3-5
3.4 Anatomy of a research proposal — 3-8
    3.4.1 Case for support — 3-8
    3.4.2 Pathway to impact — 3-11
3.5 Budgeting — 3-12
3.6 The funding process — 3-14

3.7	Summary	3-17
3.8	Further study	3-17
3.9	Suggested reading	3-18
	References	3-18

4 Presenting 4-1

4.1	Introduction	4-1
4.2	A three-way approach	4-2
	4.2.1 Developing your narrative	4-2
	4.2.2 Understanding your audience	4-4
	4.2.3 Managing yourself	4-7
4.3	Dealing with nerves	4-9
4.4	Rhetoric	4-10
4.5	PowerPoint	4-11
4.6	Timings	4-14
4.7	Answering questions	4-14
4.8	Poster design	4-16
4.9	Summary	4-20
4.10	Further study	4-20
4.11	Suggested reading	4-21
	References	4-21

5 Outreach and public engagement 5-1

5.1	Introduction	5-1
5.2	Objectives, audiences, and formats	5-3
5.3	Different publics	5-5
5.4	Working with children	5-7
	5.4.1 Children in a formal environment	5-8
	5.4.2 Children in an informal environment	5-10
5.5	Different formats	5-11
5.6	Citizen science	5-14
5.7	Funding	5-15
5.8	Advertising	5-16
5.9	Evaluation	5-17
5.10	Initiative checklist	5-21
5.11	Examples of science communication initiatives	5-24

5.12	Summary	5-27
5.13	Further study	5-27
5.14	Suggested reading	5-28
	References	5-28

6 Engaging with the mass media — 6-1

6.1	Introduction	6-1
6.2	Why, when, and how to engage with the mass media	6-2
6.3	Press releases	6-3
6.4	Constructing a narrative for mass media	6-5
6.5	Television and radio interviews	6-8
6.6	Summary	6-12
6.7	Further study	6-13
6.8	Suggested reading	6-13
	References	6-14

7 Establishing an online presence — 7-1

7.1	Introduction	7-1
7.2	Blogs	7-2
7.3	Podcasts	7-5
7.4	Social media	7-7
	7.4.1 Twitter	7-7
	7.4.2 Facebook	7-9
	7.4.3 LinkedIn	7-11
	7.4.4 YouTube	7-13
	7.4.5 ResearchGate	7-13
	7.4.6 Others	7-14
7.5	Digital collaborations	7-14
7.6	Summary	7-15
7.7	Further study	7-16
7.8	Suggested reading	7-16
	References	7-17

8 Science and policy — 8-1

8.1	Introduction	8-1
8.2	How science informs policy	8-2
8.3	What you can do to inform policy	8-4

8.4	Impact from research	8-6
8.5	Summary	8-7
8.6	Further study	8-8
8.7	Suggested reading	8-8
	References	8-9

9 Other essential research skills 9-1

9.1	Introduction	9-1
9.2	Time management	9-2
9.3	Networking	9-5
9.4	Teamwork	9-7
9.5	Objective reflection	9-8
9.6	Mentoring	9-9
9.7	Career planning	9-10
9.8	Open access	9-12
9.9	Integrity and malpractice	9-14
9.10	Promoting diversity	9-16
9.11	Summary	9-16
9.12	Further study	9-17
9.13	Suggested reading	9-18
	References	9-18

Second edition preface

In the three years since the first edition of this book was published, we have received numerous helpful comments from various parties (colleagues, students, etc), suggesting additional material that may further help the reader. Science communication, as both an academic discipline and a general practice, has also continued to advance during this time, and in carefully re-reading and analysing the first edition of this book, we realised that there was much more that we could offer.

Chapter 1 (Introduction) now provides a clearer distinction between the two aspects of science communication that are discussed in this book: that which is aimed at engaging scientists (inward-facing) and that which is aimed at engaging non-scientists (outward-facing). The section on the development of science communication has also been removed from chapter 1, and a more considered introduction to science communication as an academic discipline is instead now provided in chapter 5 (Outreach and public engagement).

Some of the most extensive changes occur in chapter 5, which has been re-structured and expanded. This chapter now includes a detailed discussion of how (and why) to engage with diverse audiences, especially those that have been traditionally underserved and under-heard by both science and science communication. Several specific examples of successful outward-facing science communication initiatives are also provided, including an exploration of why they have been so effective in engaging with non-scientists.

As might be expected, chapter 7 (Establishing an online presence) needed to be updated significantly in order to reflect rapid changes in the digital landscape. Updating the various weblinks that were referenced in this chapter also prompted us to re-consider the way in which the references were presented in the first edition, and these have now been updated and reformatted for convenience.

Throughout the book, we have included more advice about how to be an ethical and rigorous scientist, as well as a successful one, including a discussion of the need to be inclusive and to encourage and celebrate diversity. This culminates in a re-worked chapter 9 (Other essential research skills), which highlights a range of other key skills and considerations relevant to the contemporary scientist.

The cartoons of Paul Dickens were many readers' highlight from the first edition, and we would like to thank Paul for both creating new cartoons and updating others to better reflect the inclusivity that we hope this book will foster. Thanks also to the IoP Publishing Team for standardising all of the images in this book so that they look more appealing.

We hope that you enjoy this second edition, and look forward to receiving more of your thoughts and helpful comments in due course.

Sam Illingworth & Grant Allen
Manchester, October 2019

Acknowledgements

This book is the result of two years of hard work between the two of us, but there are many people who have contributed directly or indirectly through discussions and the experiences they have offered to us. We would like to thank everyone that has ever sat through one of our lectures, listened to one of our talks or put up with one of our rants. Thank you to our scientific colleagues for the innovation, inspiration, and at times perspiration that was necessary for us to shape our ideas.

Thank you also to our students and those we have met at the European Geosciences Union conferences for providing us with feedback and insight during the developmental phase of this book. We would especially like to thank Farrukh Mehmood Shahid, Alexander Garrow, and Jack Richard Varley for their help.

We would also like to thank Leigh Jenkins and the team at IOP Publishing for their help in preparing this book for publication. A big thanks also to Paul Dickens for the wonderful cartoons that appear throughout the book; we really think they help to illustrate some of points that we make and the issues that we raise. Special thanks must also be given to the two anonymous reviewers, whose comments and constructive criticisms helped to mould this book, ensuring that it was consistent and effective in its message.

Author biographies

Sam Illingworth

Dr Sam Illingworth is a Senior Lecturer in Science Communication at The University of Western Australia, with a background in the atmospheric sciences and expertise in public engagement and outreach. His current research involves using poetry and games to engender meaningful dialogue between scientists and non-scientists, for which he has secured over £250 000 in funding from a range of external funding bodies. He has an MA in Higher Education and is a Senior Fellow of the UK's Higher Education Academy, with over 50 peer-reviewed publications in high-impact journals, and is the chief executive editor of the journal *Geoscience Communication*. Sam writes several successful blogs (100 000+ readers per year) and over the last five years he has directly engaged with over 30 000 non-scientists, developing and delivering a variety of different science communication initiatives, ranging from community workshops and classroom visits to poetry performances and SciArt exhibitions. He has been an invited keynote speaker at dozens of international conferences and symposia, and has provided science communication training for over 3000 scientists. You can find out more about Sam and his work by visiting his website www.samillingworth.com, and connect to him via Twitter @samillingworth.

Grant Allen

Grant Allen is a Professor of Atmospheric Physics at the University of Manchester and currently the Director for the Environmental Science degree programme there. His research interests include trace gas measurement methods and remote sensing, especially from aircraft and unmanned aerial vehicles. Recent interests concern methods and technologies for the quantification of greenhouse gas emissions. After graduating with a PhD related to satellite remote sensing at the University of Leicester in 2005, Grant was a postdoctoral research associate at the University of Manchester, investigating tropical convection and pollution transport. This was followed by a fully funded research fellowship, leading to tenure at the University of Manchester in 2011. At the time of publication, Grant has received over £2 million in funding from the Natural Environment Research Council, United Nations, UK Government, and UK Environment Agency, for projects related to these themes. He has submitted over 40 grant proposals, with over 80 peer-reviewed publications in high-impact journals, and delivered over 100 academic conference presentations and public lectures. In 2012, he was awarded a Royal Society Westminster Pairing Fellowship to understand the science–policy interface. He is an editor for several journals and contributes to a range of scientific strategy advisory committees. Grant has also featured in many popular science documentaries and has been interviewed live on BBC and Sky News channels discussing topics from volcanic eruptions to flooding. He has also taken part in over 40 radio interviews and provided expert comment for many hundreds of newspaper articles relating to air quality and climate.

Chapter 1

Introduction

What is it that we human beings ultimately depend on? We depend on our words. We are suspended in language. Our task is to communicate experience and ideas to others.

—Niels Bohr

1.1 Introduction

As scientists we are taught the skills and techniques which enable us to perform a range of extremely complex tasks, from detecting neutrinos to modelling future climate change scenarios. Despite this, very few of us are ever trained how to effectively communicate our research, or are asked to reflect on why doing so is important. We typically work in an environment where we are told that we must 'publish or perish', and training in communication often relies on a baptism of fire for the early-career scientist. Such an approach means that some scientists still treat presenting their research at scientific conferences as being a necessary evil, while others view communicating with non-scientists as something akin to root canal surgery without an anaesthetic. The reality is that in order to be a successful (and impactful) scientist, we must be able to communicate effectively and confidently to a wide variety of audiences, using a range of different media.

For the purposes of this book we have split science communication into two broad categories:

1. **Inward-facing.** That which involves communicating to other scientists through peer-reviewed publications, grant proposals, and conference presentations, etc.
2. **Outward-facing.** That which involves working with non-scientists to both communicate our research more widely and to help diversify and broaden scientific discourse.

As scientists, the personal benefits of being able to effectively communicate our research inwardly (to a scientific audience) are relatively self-explanatory, with regards to both career progression and the accumulation of accolades. However, at times some may question what the purpose and/or benefit is of communicating our research outwardly, both to and with non-scientists. In 2017, University College London published a report into how engagement with science can be used to promote and develop social justice, stating that [1, p 2]:

> Scientific advances mean that people will need to be increasingly STEM-literate if they are to be active citizens who can have a say in society.

As practitioners of STEM (Science, Technology, Engineering, and Maths), the opportunity to create a more inclusive society through better science engagement and participation is a pretty compelling reason why we, as scientists, have a responsibility to work with non-scientists in this way.

The purpose of this book is to provide guidance on how to be more effective at both inward-facing and outward-facing science communication. In addition to you becoming a more successful scientist, giving equal consideration to improving your communication in both directions will help to make you a more useful one as well.

1.2 How to use this book

Following the Introduction, this book is split into eight chapters, each of which provides guidance for different aspects of inward-facing and outward-facing science communication. Each of these chapters presents an overview of the topic, drawn from both the literature, and also our own experiences and insights as successful academics and scientists. In each of the chapters there are also a number of exercises for you to reflect on and put into practice what is discussed, as well as further study and additional readings that are suggested to improve your knowledge and understanding.

Chapter 2 is dedicated to publishing work in academic journals. Writing your research in a format that is suitable for peer review is an essential skill for any scientist, and this chapter provides advice for how to do this effectively. An overview of the peer review process and advice on how to review the work of other scientists is also included.

Chapter 3 provides an introduction to getting your research funded, guiding you through the typical application process for proposals and also breaking down what makes for a good, and fundable, idea. A discussion of budgeting, pathways to impact, and the funding process itself is also discussed.

Chapter 4 is dedicated to improving your presentational skills, across a variety of media, for a mainly scientific audience. This chapter contains advice on how to structure your presentations, overcome nerves, and make use of rhetoric. It also provides advice on answering questions, designing scientific posters, and managing the associated timings and logistics.

Chapter 5 is focussed on outward-facing science communication, and is centred around developing and delivering outreach and public engagement initiatives for a variety of different audiences. A consideration of the various publics and formats that you may encounter is presented, alongside specific advice for working with children, and for funding, advertising, and evaluating your initiatives.

Chapter 6 presents information and advice for dealing with the mass media. This includes a consideration of how to construct appropriate narratives, how to write effective press reviews, and how to prepare for appearances on radio and television.

Chapter 7 will help you to establish an online presence, and craft a unique and useful digital footprint. Blogs, podcasts, and other forms of social media are all introduced and discussed with reference to sharing scientific research, and advice for dealing with Internet trolls and other potential difficulties and distractions is also discussed in-depth.

Chapter 8 introduces the topic of science and policy, outlining how the two interact, and providing information on what you can do to better inform policy through your scientific research.

Finally, **chapter 9** catalogues a series of other essential research skills, such as time management and networking, and outlines how and why you can develop each of these. This chapter also discusses in detail what it means to be an ethical scientist, and provides assistance for planning your future career, either within, or away from academia.

Whether you are an undergraduate scientist embarking on your first steps into the exciting world of scientific research, or a professor with dozens of years of experience, this book contains guidance and advice that will be (at least in part, if not completely) relevant to you. Being a scientist is an incredibly rewarding and enjoyable experience, but it can also be a testing and difficult one as well. We hope that this book acts as a handbook for improving your ability to communicate effectively; and that in doing so it is also a practical guide to surviving, and thriving, as a scientist in the 21st century.

> **Exercise: what do you want to improve?**
>
> Write down three personal aims for being a more effective scientist. These aims should be SMART, i.e. Specific, Measurable, Achievable, Relevant, and Time-Bound. For example, 'Write more publications' is not a SMART target, whereas 'Author or co-author two research articles by the end of this calendar year, in high-impact journals' is. Try to include at least one outward-facing science communication aim in your list of three.
>
> Now, turn to the table of contents and select the chapters that are most applicable to your three aims. After working your way through the appropriate sections, return to your aims and rewrite them to be more realistic. Alternatively, if you are one of those people who absolutely has to read a book in page order, and from cover to cover, then refer back to your aims as you reach the appropriate section, re-evaluating them as you work your way through the book. Keep you three aims in a location that is easily visible (e.g. on your desk or in the front of your lab book), and re-evaluate your progress in achieving these during regular intervals (e.g. once a month). Once you have achieved an aim cross it off your list, and once all three have been achieved do something nice to celebrate, then make another list and start again.

1.3 Summary

To become better scientists, we need to be more effective at both communicating our research to the scientific community, and discussing it with non-scientists. By reading this book, working through the exercises, and following the recommendations for further study you will learn the skills that enable you to do this.

1.4 Further study

The further study sections at the end of each chapter in this book are an opportunity for you to reflect on what you have learnt, and to develop some of these ideas through further reading and practice.

The further study in this chapter is designed to get you thinking more about outward-facing science communication, and the benefits that this can have for the wider society:

1. **Ask the news.** Visit the website for a news outlet of your choice, and find the 'Science' section. Is what they are reporting based on peer-reviewed scientific research? Does it appear to be logical? Would somebody without your scientific training be able to follow the story and come to the same conclusions?
2. **Ask a non-scientist.** Find a family member or a friend who is a non-scientist and ask them about their opinions of science. Can they define 'science'? Can you define it? Do they think that scientists are good communicators? Do they trust scientists, and if not, why not? Do they think that science offers a positive contribution to society?
3. **Ask yourself.** Take a moment to reflect on what it was that made you want to become a scientist in the first instance. Did you have a particularly inspiring

teacher? Were you encouraged by your family? Was it just something that you were really good at and wanted to pursue further? Keep these reflections in mind as you work your way through the rest of the book, and remind yourself of them when you next need some scientific encouragement.

1.5 Suggested reading

The Demon-Haunted World: Science as a Candle in the Dark [2] provides a convincing and passionate argument for why scientists should make science more accessible, and the consequences of a population being scientifically illiterate. Written by the American astronomer and science communicator Carl Sagan, it is a powerful reminder as to some of the responsibilities that we have as scientists. A good introduction to science communication as an academic discipline is provided by both *Introducing Science Communication: A Practical Guide* [3] and *Creative Research Communication: Theory and Practice* [4]. There are also several journals that are dedicated to the development of this subject, including *Science Communication* from SAGE Publications [5], and the *Journal of Science Communication* from SISSA Medialab [6]. However, the format and nomenclature of these journals can at times be inaccessible to scientists who are new to science communication. *Geoscience Communication* from Copernicus Publications [7] provides a more accessible entry into the field, being written primarily by and for geoscientific researchers who are interested in applying their scientific training to their outward-facing science communication initiatives (see chapter 5).

References

[1] Godec S, King H and Archer L 2017 *The Science Capital Teaching Approach: Engaging Students With Science, Promoting Social Justice* (London: University College London)
[2] Sagan C 2011 *The Demon-Haunted World: Science as a Candle in the Dark* (New York: Ballantine Books)
[3] Brake M L and Weitkamp E 2009 *Introducing Science Communication: A Practical Guide* (London: Macmillan)
[4] Wilkinson C and Weitkamp E 2016 *Creative Research Communication: Theory and Practice* (Oxford: Oxford University Press)
[5] Science Communication https://journals.sagepub.com/home/scx (Accessed 16 October 2019)
[6] Journal of Science Communication https://jcom.sissa.it/ (Accessed 16 October 2019)
[7] Geoscience Communication https://geoscience-communication.net/ (Accessed 16 October 2019)

IOP Publishing

Effective Science Communication (Second Edition)
A practical guide to surviving as a scientist

Sam Illingworth and Grant Allen

Chapter 2

Publishing work in academic journals

If I have seen further it is by standing on the shoulders of giants.
—Isaac Newton

2.1 Introduction

This chapter offers some advice on how to publish a peer-reviewed scientific paper, laying out a framework from conception to publication. Based on our personal experience as an editor, a reviewer, and an author, this chapter will provide practical advice and guide you through the process of publication in a typical modern scientific journal. We shall track a paper's journey—from deciding on when you have something to offer to science, to identifying an appropriate journal to publish in; and then how to navigate the peer review process and ensure your published paper reaches its target audience. The advice in this chapter is especially relevant to those embarking on preparing their first scientific paper, but it may also offer some helpful insight to maximise a paper's academic impact and reach for those with some existing experience.

Writing and publishing peer-reviewed academic journal articles remains the principal way that scientists communicate their research widely among other researchers. Unlike other scientific communication methods, which may have a greater value in reaching wider audiences, peer-reviewed journal publications in reputable journals represent a time-honoured gold standard in academic rigour, and provide a permanent record of contributions to humanity's body of scientific knowledge. This is because the checks and balances of the peer review and editorial process serve as an important quality control on the accuracy and rigour of the work presented to a journal, and serve to keep science honest in the face of constructive criticism and independent oversight. On final publication, science can be reassured that an individual's or team's work has been carefully and reasonably vetted by independent experts familiar with a particular field, and that a published article has addressed any reasonable concerns. Peer review (like any human endeavour) is not perfect (as we shall discuss later), but it does represent the best system we know of in science to ensure accountability and scrutiny. First and foremost, peer review is intended as a constructive process, and should be approached in this manner by both reviewer and author; though that may not lessen the sense of anxiety some may feel when they receive reviews relating to their latest submission.

Bringing a paper to publication can be a daunting but also very enjoyable and rewarding experience. It is our duty as scientists to publish our work and bring it to the attention of others who may learn from it. The quote at the head of this chapter encapsulates the engine of scientific progress—all of our current knowledge and teaching emanates from those that have previously published their findings for us to learn from. We build on each other's work and move our own forward incrementally. The record that is our academic literature ensures that our work is forever open to scrutiny such that it may be refined, disputed, or reinforced with time, in the light of future understanding and effort. In the modern scientific world, the number of journals (and the number of scientists) has been growing exponentially. This has many strengths but also some weaknesses, as we shall discuss later. However, the process and the end result remain the same—to record knowledge and take it further. Let us now explore how you can make the most of this process in your work.

2.2 Scoping your deliverables

The word 'deliverables' is well-used in academia these days, implicitly commercialising science by its virtue as a term borrowed from the world of business. However, it serves its descriptive purpose. A journal article or paper is an academic output that contains a deliverable, or deliverables. These deliverables are the key conclusions of the paper that represent new pieces of information not previously known to science. They could represent enormous leaps forward in fundamental knowledge, or they could represent incremental advances or facts about the Universe (or anything in it). The relative importance and the scientific field of those advances may dictate the journal you choose to submit to (see section 2.3) but the fact remains that any paper must contain some new contribution to knowledge, no matter how large or small. This simple requirement is one of the first aspects that an editor or reviewer will look for and be asked to comment on concerning any submitted paper. Therefore, the first step in writing a paper is to recognise when you have something useful to say. All that follows from that point, to prepare a paper for submission, concerns framing the deliverables, to provide clear evidence and explanation so that others can be confident in your conclusions. A key tenet of the scientific process is that others may reach the same result following the same methods using the same dataset. This is the principle of repeatability and it goes to the heart and the origins of the scientific method. While the interpreted conclusions made in any paper may be tainted by subjective opinion, the data and results should be absolute according to a transparent and rigorous method. As datasets and methods become increasingly complex, because of the sheer volumes of data from modern instrumentation, and the analytical power of computers and algorithms, the principles of repeatability and transparency are getting increasingly harder to ensure. This modern problem has been recognised recently by many of the leading journals [1], which are leading a campaign on best methodological practice, and open access (OA) and archiving of data (and metadata). After a string of high-profile academic malpractice cases [2], better processes to challenge and report poor practice (and outright malpractice) are also being introduced. Time spent becoming familiar with best practice, and reflecting on how to embed the principle of repeatability in your work, will pay enormous dividends, both to yourself and to science, as those that follow your work will be able to have more confidence and respect for the outputs you make.

From experience of supervising students and researchers, and as former students ourselves, it is not always easy to recognise when a critical threshold has been reached in the context of having something 'useful' to publish. For some it may be easier than others; for example, if the deliverables were planned in advance of fairly routine analytical work that was carried out. But in many instances, science moves forward by stumbling on some unexpected new advance as a result of working on something else. At this point it is important to take a step back, explore what you have already found, and decide on three things before deciding on whether to write a paper:
1. Is what you have found so far a scientific deliverable that others should hear about?
2. If so, does the work done so far represent enough information, data, or explanation to provide a coherent and substantial narrative from which to inform others of that advancement?
3. If so, could that work be written up as a paper now, or could further work provide additional deliverables within the scope of the intended article?

Some of the words in the above list are necessarily subjective or vague. This is because every discipline and every piece of work is different. However, choosing the right point in the course of your work at which to publish is a skill that you learn to develop with practice. The difference between a mediocre paper and a truly ground-breaking one could be as simple as gauging when there is a neat package of work to create a clear and full story, as opposed to publishing as soon as there is something new to say. However, delaying publication while waiting for new results is a risk that involves making educated decisions about the future direction and timescales of the work you might be engaged in. If in doubt, the less risky option may be to publish as soon as possible. Prior to journal publication, an alternative may be to consider publishing a pre-print of your article (see, e.g. [3]). Pre-prints are an excellent way to get scientific information out to a wide audience quickly, especially for work that may relate to something of high immediate public or scientific interest. Pre-prints also offer an opportunity to get alternative feedback from peers outside of a formalised journal peer review process. Some choose to do this because of the typically long timescales involved between submission and publication for many journals. However, because pre-prints are not peer-reviewed, they are often considered as 'grey literature', meaning that it is not best practice to cite such work in more formal publications.

Implicit in point 3 above, is knowing where and when to stop, where to draw the line under a body of work and when to present it to others. Of equal importance to knowing when you have something to say, is to know when not to say too much. This is absolutely not about holding anything back, but it is about knowing how to properly scope an article. It is often just as hard for new researchers to know when to stop (and publish) as it is to know when there is something useful to say. There is always a temptation to keep going. But remember that science is never-ending (or at least we hope); it is therefore an important career skill to recognise when to compartmentalise your work and bring it to fruition in the form of a scientific article. This is not to say that you should abandon a thread of research once you have a good deliverable on a given subject, but a paper should be relatively self-contained and address deliverables in the context of the title of the paper. It may not make sense to produce a paper for every new deliverable where there are variations on a theme for example, but it is important to know when to stop before a paper runs the risk of becoming unwieldy.

Something to avoid is falling into the trap of believing that simply publishing more papers is always better for your career. In a world where the length of a scientist's publication record is a cursory symbol of academic success, it is tempting to add quantity to that list at the expense of quality. In reality, both quality and quantity matter, and quality often matters far more. A long string of papers that have just made it past the threshold of publication acceptability, in a low impact journal with only meagre deliverables, may be meaningless if no one chooses to read the paper or cite it in their own work. Increasingly, academic reach and success is rated in terms of the number of citations a paper may receive (see section 2.7). A high-quality paper with important and useful deliverables on a well-scoped theme may attract a higher readership and hence more citations. Such an output is far more important for your publication record and self-esteem, and more importantly, far more useful to science.

In essence, deciding on when and what to publish should be an informed balance between the completeness and complexity of understanding from research already done, and the promise and direct relevance of what may still come from further work. Scoping out your potential scientific deliverables as you go is a useful way to keep this balance in focus.

Exercise: scope out your deliverables

This exercise will help you to think about your own research and to scope which aspects may lead to scientific papers now or in the future.
1. If you are currently carrying out a research project, list what aspects of your research (past or future) may represent original contributions to scientific understanding.
2. From your list, group any aspects into specific themes.
3. For each of those themes, think about whether they would be best presented in a stand-alone paper, or whether they could be grouped into sections of one

umbrella paper. Remember, more papers are not necessarily better—but good scoping is.
4. For each of those themes, think about what work you may still need to do to fully address them. If more work is required, do you still have important deliverables to publish on the other themes at present?

2.3 Choosing a journal

Once you have decided you have something useful to say, the next task is to decide on which scientific journal will best speak for you and your research. A journal is the medium through which we permanently record and communicate our research in its most complete form. Choosing a journal is analogous to deciding whether to present work at a large but general, or small but specialist, conference—each has relative strengths and weaknesses depending on the scope of the work.

The scientific journal landscape is vast, and growing. Virtually all scientists (and many non-scientists) will have heard of, and read, publications such as *Nature* or *Science*. But only a small number of specialist researchers may regularly read the *Journal of Waste Management*, for example. The relative reach and specialism of different journals reflects the scope and wider import of the articles that each publishes. For example, *Nature*'s readership may well not be too interested in the finer details of anaerobic digestion of organic waste in landfills, whereas the *Journal of Waste Management*'s readership may be surprised to read an article about new predictions of global climate catastrophe due to greenhouse gas emissions from landfill. But each article has its rightful place, and its attentive audience. Therefore, a key aspect when choosing a journal is to think about the scope of your conclusions and which group of people will be best served by hearing them. The journal you choose should then be analogous to selecting the loudest microphone positioned in the most appropriate room of people.

One of the metrics of a journal's reach is its impact factor (IF). This is defined as the ratio of total citations for the journal in some time period (usually two years) to the number of articles published in the same time period. These are routinely published by academic journals (usually on the homepage of their website) and in league tables produced by various organisations that can be found easily in any Internet search. The higher a journal's IF, the more impact that journal's articles can be assumed to have by virtue of the fact that others are referring to work published there. So generally, the IF is a proxy for the relative importance of the journal within its field. In reality, this IF ratio reflects both the magnitude of the journal's readership and the quality (and scope) of the articles it chooses to publish. Thus, many of the highest IF journals such as *Nature* and *Science* publish only the highest quality of articles with broad and societally-important themes that carry interest to a wide (and even non-scientific) audience. Such wide-ranging articles are naturally positioned to be more readily cited by others, whereas more technical articles in specialist journals may be less obviously cited. Put simply, it is important to attempt to publish in high IF journals where possible to maximise reach and impact. But your choice of journal may be limited by the scope and import of your subject matter and its conclusions. In all cases, your choice should be about reaching the attention of an appropriate audience for your work.

As briefly discussed above, one of the metrics of success for academic outputs concerns the IF of the journals we publish in and how often our individual articles are cited. However, not all articles are suitable for the highest IF journals. A specialist technical article may be best suited to a specialist journal with a smaller readership and a lower IF. In other words, the choice of journal should reflect both the IF and its scope, and it is always important to target a journal with a high IF for the specific field of interest.

A recent proliferation of academic journals has accompanied the growth of science as a global commodity, and the profits that publishing companies may stand to make from publication fees from a growing population of scientists. The era of digital online publishing has also removed much of the cost barrier in starting up new publishing enterprises that typically accompanied the print media of the past. Some of these new journals have become highly successful and respected and their IF has grown dramatically. But beware. Recent research has shown that this system is ripe for abuse (and is being abused widely) [4, 5]. Some of the most unscrupulous of these 'predatory' journals pay lip service to the quality of the peer review process (or bypass it altogether) [6], taking publication fees for profit at the expense of academic quality and rigour. Unfortunately, some career-hungry scientists, jaded by rejection or heavy revision as part of the peer review process in more prestigious journals, have elected to feed the growth of these predatory journals to add bulk to their publication record. A growing awareness of these predatory journals and regularly updated blacklists does mean that this problem is slowly being dealt with within the academic community. However, for the uninitiated (e.g. employers outside of academia), a publication record can still, unfortunately, be taken at face value. Here, the advice is simple—do not publish in such journals and always check that a journal that you are considering is not included on any blacklist. The easy ride they may appear to offer is at the expense of the quality of your published work, and represents a blemish on any academic record. Publishing in these journals is akin to buying a bogus qualification and is considered by most to be a crime against academia.

To find a genuine and appropriate journal for your article, a good place to start is to read recent issues of journals that others in your field have published in, to get a sense of both their scope and quality. When conducting any literature review you may perform prior to embarking on a research project, or when writing the introductory and discussion sections for your own paper, you should naturally become familiar with a range of appropriate journals relevant to the context of your work. When you have compiled such a list, visit the journals' websites to investigate their IF and read about their aims and scope. Your choice of journal should then be concerned with which of them will give you the greatest reach based on the journal's IF, scope, and likely audience.

A further consideration also concerns whether the journal has an OA model. OA refers to a funding model under which published articles are freely available to anyone who has access to the Internet; as opposed to previous closed access models, where individuals or institutions (e.g. universities or libraries) may have had to pay a regular subscription or per-article access charge [7]. This funding model relies on the payment of publication charges by the author at the time of publication and/or on contributions from institutions or other funding agencies. A recent trend in enhancing the visibility of research and its accessibility to wider non-fee-paying audiences means that the popularity of OA journals has grown markedly. Furthermore, in many countries where research has been commissioned with public money, it has become a requirement of funding to publish in OA journals such that the public has free access to the results of work that has been commissioned with their money. Given this trend, and the natural enhanced visibility that OA offers, it would be advisable to explore OA journals when making your choice of which journal to publish in. A further discussion of the OA model, and why it should be supported, is given in chapter 9.

2.4 Writing and manuscript preparation

In this section, we discuss some tips for journal publication preparation. However, in this book we do not offer detailed advice on scientific writing style, suffice it to say that the best way to learn is to read widely around your subject, to learn from the best practice of others, and then to put pen to paper (or rather finger to keyboard) yourself and be prepared to iterate tirelessly. Other published material offers some excellent and detailed advice on scientific writing style (see, e.g., [8]).

Before moving on to discuss preparing manuscripts for publication, one piece of advice on style and writing we would offer is to thoroughly check your manuscript prior to submission for typographical and grammatical errors, and to get your work proof-read by a technical writer or proof reader if you have this resource available to you. A well-prepared and clear text will go far with any reviewer. Similarly, nothing gets reviewers worked up more than having to write out a long list of technical errors such as grammatical and typographical mistakes. If you put your reviewers in a bad frame of mind over this avoidable aspect, then they may be less objective about your technical presentation, siding with a presumption of sloppiness. Put simply, typographical and grammatical errors push the burden of correction onto reviewers and editors who simply shouldn't have to take on such a role in what is ultimately an unpaid duty to science, and which relies on the good will of busy people. And most importantly, remember that papers can, and frequently do, get rejected if writing style is so poor as to affect the ease of reading, and therefore the clarity, of submitted work. This may seem like common sense, but ignorance of this is unfortunately all too common.

Once you have decided on a journal to submit your paper to, it is useful to thoroughly read through any author guidance material that is available. This is usually accessible from the journal's website under a tab such as 'Guidance to Authors' or similar. This guidance will typically contain rules and advice on how your submitted article should be formatted or prepared, and it may detail how items such as figures, tables, and references should be provided.

Most journals will accept a word-processed document with embedded figures and tables for any initial review phase. But you can save yourself a lot of time later on if you have prepared all material in any required format in advance. This is especially true for any figures, photographs, or illustrations which usually have strict file formats; for example, encapsulated post-script for data plots and raster graphic formats for any illustrations. As many scientific analysis software packages have the ability to output graphics in a range or file formats, it could save you hours (if not days) of time to prepare all outputs in the journal-required format at an earlier stage, as journals typically require separate files for each figure during the final stages of copy-editing. Other guidance to look out for includes any word or page limits, as some journals that take letter-style articles usually stipulate strict restrictions. It is also worth considering the graphics and colour schemes that are used in your article, as these can really make a big difference to how well received your article is by both readers and reviewers. An excellent resource for making your images best represent the underlying data is the Climate Lab Book website run by Professor Ed Hawkins [9].

Before submitting an article to a journal, make sure that any co-authors have read and commented on your draft, and ideally pass it to someone uninvolved with the work (perhaps someone else in your research group) to provide a fresh pair of eyes and comment on how accessible the narrative is to someone at the edge of your field.

One of the aspects of submission that most often gets overlooked is the preparation of an abstract. While every journal will have their specific requirements for what an abstract should look like, ultimately this should be seen as an opportunity to highlight the key deliverables of your paper and maximise their impact. Abstracts are often used by scientists to determine if the rest of the paper is worth reading (and potentially citing in future research), and so it is vital that you signpost the key findings of your work in an accessible and well-structured format.

Rather than treating your abstract as an annoying necessity, see it as an opportunity to succinctly communicate the importance of your work, and the first step in helping it to get your research properly recognised (see section 2.7).

Finally, give due consideration to the cover letter that accompanies your submitted manuscript. Most journals will insist that you provide one, but it is good practice to always prepare one even for those that do not. This cover letter should be addressed to the editorial board of the journal, and it is your opportunity to explicitly explain what your deliverables are in plain language, and why they are relevant to this journal and its readership. Cover letters are often glossed over, but they are an opportunity for you to demonstrate to the editors of the journal exactly why your article should be considered for publication, especially if your research is particularly novel or on the fringes of what might be considered as suitable for that journal's aims and scope. Oftentimes a good cover letter can prove the difference between a paper being submitted for peer review or being rejected by the editor before it reaches this stage.

2.5 The peer review process

In all reputable journals, after you have submitted an article (usually online) it will first be assigned to a handling associate editor (AE) for the journal, who will typically have a reasonable (but not necessarily always high) level of expertise in the field of research with which your article is concerned. It is possible to examine or request a list of AEs for your chosen journal, and to make a recommendation based on your reading of their expertise in a covering letter to the journal which accompanies your paper on submission. However, the assignment of an AE is ultimately made by the publishing staff or a senior editor. The AE will preside over the peer review process of your article and they will make any final decision on suitability for publication.

On first receiving an article to handle, the AE may make an initial judgement on whether the subject matter of the article fits the scope of the journal and whether the initial manuscript is suitable to send on to reviewers, taking into consideration the arguments that you have laid out in your cover letter. It is not uncommon with some journals to receive guidance from the AE prior to further peer review at this stage. Any comments from the AE at this point should be carefully examined and answered to allow your paper to progress further. The AE will then select and invite several expert reviewers to comment on your paper. This phase can sometimes take several weeks while reviewers accept or decline invitations to review, requiring the AE to find suitable alternatives. It is important to remember that peer review for journals is almost universally a service that other researchers provide for free in their spare time. This can sometimes make it difficult for journals to solicit suitable reviewers quickly. Again, it is possible for you to suggest sensible reviewers to the AE. However, it is important to avoid bias in your choice and only suggest expert reviewers that are not linked to your work or organisation, as an AE will take a dim view of any attempt to undermine or subvert the quality of the peer review process. The AE may then choose to invite one or more of your suggestions but this is entirely at their discretion.

Once expert reviewers have been assigned (typically two or more), the reviewers may take several weeks to complete their review of the submitted paper. They will be asked to comment on the suitability of the subject matter for the journal, to report on the quality and importance of the work, and to discuss any technical points where there may be cause for question or concern. They may also be asked to comment on the quality of the figures and tables, and to list any technical errors such as typographical and grammatical mistakes. Finally, the reviewers will be asked to make a recommendation for publication, and sometimes they may be invited to submit a score against a set of criteria. This score or recommendation may not be made visible to you, and it is important to remember that any reviewers' comments and recommendations are advisory to the AE and do not represent a decision on publication, which is usually the decision of the AE alone.

After all of the solicited reviews have been received by the AE, they will contact you and provide the reviews along with their decision on publication. Rarely, an article may be accepted outright in its current form, with no further modification necessary. However, most often reviews will be returned to you with some guidance from the AE on how to proceed in their opinion. This can range from outright rejection, where the article has been deemed to be out of scope with the journal or of insufficient quality to be improvable for publication, to a suggestion for revision based on the comments of the reviewers. Suggestions for revision typically take the form of either a major revision, where significant further technical and/or presentational work may be required, or a minor revision where further clarity or graphical improvements may be required, for example.

THE PEER REVIEW

On receiving the AE's decision and guidance, you will then be given several weeks to prepare a response to the reviewers' comments, provided that your paper has not been rejected outright. In your response, it is always courteous to thank the reviewers for their comments and to briefly summarise what you see as the salient points of their collective reviews, before continuing to address each review and each reviewer's comments in turn and in order. This logical sequence will make it easier for the AE to follow your response.

It is useful to be mindful that the vast majority of reviewers are busy people that have chosen to give up their time for free to help you as constructively as possible. They have been selected to review your work as recognised experts in the field of your own paper. For the most part, peer review is a highly valuable and helpful process. But your skill as an author is in recognising which reviewer comments are accurate and helpful and which may be in error. It is not uncommon for a new author to feel that everything a reviewer says is correct and that you may always be wrong. Self-doubt and reflection, and a willingness to revisit ideas and conclusions in the light of peer review, are an important and valuable quality in the scientific process, but not at the expense of the truth. When reading and reacting to reviewer comments, you must face up to those comments honestly; addressing them where they raise important points, but also robustly defending your work where there may be misunderstanding.

Most importantly, be sure to constructively address and respond to all of the comments made, whether that is to agree with the comment or suggestion, or to discuss or argue your case. In all cases, you should detail where and why you have made changes to your paper as a result of any of the reviewers' comments, and describe how your changes have addressed the points raised. Remember that your reviewers are objectively working to help you improve the presentation and accuracy of your work, so if you see that a reviewer has misunderstood something, try to think about why they have misunderstood it and attempt to clarify any parts of the narrative or material that could lead future readers to similar misconceptions.

Very rarely, you may receive what can only be described as an unconstructive review. This class of review may take a very negative tone and may make unsubstantiated comments with no link to the content of your paper. Our advice here would be to treat such reviews with the contempt they deserve, if it is obvious that a reviewer has made no attempt to rationalise their comments. Make the point that you have no case to address if you have not been given cause to. Your AE will almost certainly have already identified such a review as being worthless to the decision process and may have selected additional reviewers to provide more objective comments. This said, be sure to differentiate between negative but objective reviews that do make specific comments linked explicitly to your work, and reviews that are generally ignorant of its content. In other words, never mistake a review you just don't like, but is genuinely raising comments about the content of your work, with a review that makes no substantiated comment at all. The former requires a detailed response, while the latter requires very little. Thankfully, such reviews remain extremely rare. But if you are unlucky enough to receive one, remember that the checks and balances of both the other reviewers and your AE, coupled with your right to respond, all serve to optimise the quality of the peer review process.

On submitting your revised paper (if required to do so by the AE), your paper may be sent back to the original reviewers for further comment, or it may now be deemed acceptable for publication without further review. The AE will usually make any decision for further review based on your responses to the reviewer comments, taking into account whether you have satisfactorily addressed them all in your response, and that any appropriate and necessary revisions to the paper have been made. In some circumstances, this iterative process can happen several times before the AE is satisfied that all aspects have been sufficiently addressed and that final publication can proceed.

Finally, and often several months after you submitted your original article, you may receive an email from the AE (or an editorial assistant) informing you that your paper has been accepted for final publication subject to copy-editing and final typesetting. This is a heart-warming moment for any researcher at any stage in their career and it is a cause for celebration. At this stage, you can be rest assured that you have made a contribution to knowledge and that your work will be recorded for future generations to build on. You now 'stand on the shoulders of giants'.

2.6 Reviewing papers

Very shortly into any published research career, you may find that you are invited to review papers yourself. Such an invitation is an honour and reflects your growing reputation in your field of research; it also represents an altruistic duty that all researchers rely upon each other to perform. Without the unpaid work that expert reviewers and AEs do for journals, the peer review process would grind to a halt. And without that process, the quality of published work and the rigour of scientific endeavour would suffer immeasurably. As often very busy people, some scientific researchers fail to see the value to themselves in peer-reviewing the work of others, but without that contribution from each and every one of us, their own work would ultimately be devalued. As a rule of thumb, we try to accept invitations to review at least as many papers as we publish, although we accept that this might not always be possible. As an emerging researcher, taking part in this process is of great value, affording you the opportunity to learn something new at the same time as critically evaluating others' work, thereby helping you to develop your own writing style and technique for future submissions.

Accepting an invitation to review for the first time can be as daunting as publishing your own work; it is a great responsibility and as such requires time and effort. After all, you are being asked to provide a judgement on work that another researcher or team of researchers has spent a significant amount of time preparing. Your job is to remain objective, constructive and honest, and to only accept a review if you feel that you are reasonably well qualified to comment on the subject matter of the paper in question. If there are areas of the work that you cannot comment on, be sure to make this clear to the AE and the author(s) in your review.

A useful way to learn about writing reviews is to read the reviews that others have published online. Many OA journals also have an open peer review process, where the reviews of papers and the responses from authors are published publicly during a discussion phase. The review and discussion boards of some more contentious publications make for some very interesting reading indeed, and can be just as informative as the final article itself. This open discussion is an exciting new addition to the modern era of scientific publication but unfortunately it can also (very rarely) attract some of the more negative aspects of social media such as anonymous trolling (see chapter 7). Again, it is important to remember here that the checks and balances of the peer review process prevails in virtually all instances. Another possible criticism with this model of peer review is that it means that a draft of the paper gets published online prior to peer review. While many may argue that this 'pre-publication' adds to the noise of science, it is also a fair and efficient way of making sure that science is conducted in as open and accountable a manner as possible.

Exercise: preparing for peer review

By thinking like a reviewer when reading, you will remain more objective about others' work and conclusions, and improve your own paper writing by putting yourself in the position of someone unfamiliar with your work.
1. Choose a paper that you would like to read and that you have not read before (ideally something relevant to your current research).
2. Write a review of that paper as if you have been asked to review it by an AE. Structure your review as follows: (a) summarise the conclusions of the paper and your overall opinion of its importance and quality; (b) list your specific comments about aspects of the work that are not clear to you or that you feel may be incorrect, asking specific questions as if you were asking the author; (c) list any technical comments such as typographical or grammatical mistakes (though there shouldn't be any of these in a published article!).
3. Read over and reflect on your review. If you received this review as an author, how would you interpret it? Are all of the points that you have made objective and clear? Does your review help the AE to make an informed decision with regards to publication?

2.7 Citations and metrics

In section 2.2, we discussed how the quality of published work is perhaps more important than the quantity of publications in your record. One of the more explicit measures of your work's quality and its importance to science is the number of citations your paper may receive. Clearly, your choice of journal and its IF are a major influence on its exposure and therefore the chance that others will read and build on (and cite) your work. However, there are other ways to raise your work's profile in your field. These concern how you advertise your work to others, and there is much that you can do to bring it to the attention of relevant researchers beyond simply relying on the journal and random Internet searches that others may perform when conducting a literature review of their own. While some researchers do still monitor and read every issue of their favourite scientific journals, an increasing number of researchers rely instead on filtered journal alerts or occasional literature searches through Internet search engines. With this in mind, there are several ways to make sure that your work appears on the radar of those people that do not actively trawl the contents pages of traditional journals. The first of these is to carefully choose the search index keywords that many journals ask you to stipulate when you submit an article. You can usually specify several such keywords or phrases that summarise the subject matter or field of your paper. For example, these may be 'greenhouse gases', 'nanoparticles', or 'unmanned aerial vehicles'. Think about what words you would search for when conducting a literature review related to the content of your own paper. Try those keywords out and see if they list similar papers to your own. As discussed in section 2.4, it is also vital to ensure that your abstract succinctly communicates the key findings of your research in an informative and enticing manner. When scrolling through hundreds of research articles on a particular topic (or keyword), most researchers will use the abstract as an initial filter to determine if the study is worthy of further reading.

Other more active ways to advertise your research include presenting your recent work at conferences and to include reference to your papers in any abstracts that you submit to them. You might also consider listing your most recent papers in your email signature. And of course, it is always useful to directly email any researchers you know in your field who you think may find your paper interesting, as well as taking every opportunity to advertise your paper through all of the social media channels that you utilise. Some of these channels, such as ResearchGate, Mendeley, and LinkedIn (to name only a few), offer the ability to list your publications and form networks with other colleagues and researchers who may be automatically informed when you publish new work; a further discussion of some of these sites is given in chapter 7. Lastly, any organisational or personal websites should be kept up to date with your evolving research interests and publications.

Keeping track of your papers' citations is useful, as it will help you to recognise how your field is developing and to identify which of the papers that you have written have been the most well received. There are several conventional metrics that can be used to help formalise this process; these metrics are also often used in academic or scientific-related career promotion criteria, and as such they can be a hotly debated topic. The more common of these metrics include the 'h-index' and the 'i-10 index'. The h-index is the most common of the two and is based on a set of a researcher's most cited papers and the number of citations that this same set may have received in other publications [10]. The value of this index is calculated such that a h-index with an integer value of 'h' represents an author that has published h papers, each of which has been cited in other authors' papers at least h times (see figure 2.1). As such, this index then reflects both the number of publications and the number of citations per publication. For example, an author with 20 publications but with only 5 papers that have been cited at least 5 times, will have an h-index of 5, while an author with 20 publications, each cited at least 20 times would have an h-index of 20. Clearly, this index favours those with consistently well-cited papers, but not those with one stand-out paper and several less well-cited papers. An important point to note is that the h-index is rarely relevant when comparing the impact of researchers across different disciplines due to the differing ways in which different fields publish and cite each other's work. However, it can be an effective way to compare the impact and reach of individual researchers within more common disciplines. The most effective way to ensure that your h-index continues to rise is to publish high-quality papers and bring these to the attention of as many fellow researchers as possible. The i-10 index is a measure used solely by Google Scholar at the time of writing. This index is defined as the number of publications written by an author with at least 10 citations. As such, this is a simpler index to understand compared with the h-index, but it further amplifies any interdisciplinary comparison bias.

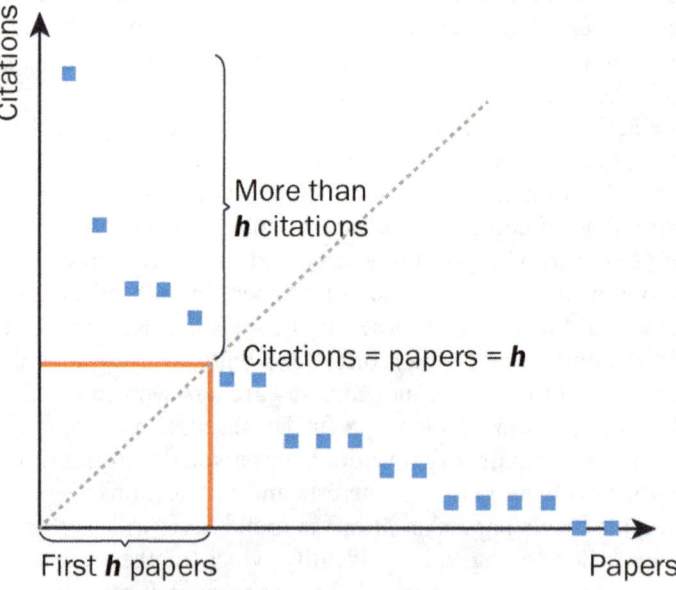

Figure 2.1. Calculation of the h-index using the number of papers attributed to a given author and the number of citations each has received. The h-index is then the maximum integer value where that number of papers has received at least the same number.

There are many more exotic variations of these citation metrics, all of which are designed to remove a source of discrepancy, such as the discipline bias or age of the author, but not all metrics are readily recognised or understood by many (see [11] and references therein for further examples). A further point is that the value of these indices may vary wildly between different citation-indexing services, and that not all organisations recognise indices calculated from some services due to the perceived impurity, or chance of false positive data included in their calculation. It is your duty to honestly examine your citation indices to check that they have been calculated correctly, using the most up-to-date information. This may require you to routinely weed out (or add) any publications or citations that may have been wrongly attributed to you in any specific indexing service.

2.8 Summary

This chapter has described the importance and necessity of publication in scientific journals as the mainstay of academic record and communication. We have briefly explored the pathway and the peer review process when submitting scientific papers to traditional academic journals. We have also offered some tips and advice on maximising the exposure and impact of your research outputs, and how to avoid the grasp of predatory journals. Finally, and as important as publishing your findings, we have discussed your service to the academic community (and science in general) by engaging actively with the peer review process.

2.9 Further study

The further study in this chapter is designed to help you think further about developing your paper writing and reviewing skills:
1. **Read an article from outside of your discipline.** Pick a scientific article from a reputable journal that lies outside of your specific area of expertise. Read that article and see how it has been constructed in terms of its structure and layout. This will help you to defocus a little on the content of the paper, allowing you instead to reflect on its style and structure. Does the paper present its research findings in an innovative way, or is it overly verbose and difficult to decipher? Are there any lessons that could be learnt for your own writing?
2. **Follow an online discussion.** Pick a journal from within your field that has an online and open peer review system. Search through some of the articles until you find one that has a particularly long comments thread (ideally one that has inspired other members of the community, other than the compulsory reviewers, to comment), and see if you agree or disagree with some of the comments that have been made. Have all of the comments been written in a constructive way, or are there instances of unprofessionalism or a lack of objectivity?
3. **Compile some metrics.** Pick a couple of very well-known scientists in your respective field, both alive and dead, and compile a list of the various metrics that would be used to rate their publication records. How do they perform? See how they correlate to similar metrics of researchers from another discipline.

2.10 Suggested reading

While targeted specifically at the atmospheric scientist, *Eloquent Science: A Practical Guide to Becoming a Better Writer, Speaker, and Atmospheric Scientist* [8] is a book that contains a wealth of generally useful guidance on many aspects of scientific writing. 'The science of scientific writing' [12] also offers some helpful advice on how to improve the quality of your scientific writing; similarly, if English is not your first language then *English for Writing Research Papers* [13] is an essential guide for helping to avoid common mistakes, and increase the readability of your work. If you are interested in developing your skills as a reviewer, then 'How to write a thorough peer review' [14] provides a straightforward methodology for constructing a considerate peer review, and also provides a useful worksheet for guidance. Finally, for readers wanting to find out more about the open peer review system, 'What is open peer review? A systematic review' [15] presents an excellent introduction.

References

[1] Tennant J P 2018 The state of the art in peer review *FEMS Microbiol. Lett.* **365** fny204
[2] Haug C J 2015 Peer-review fraud—hacking the scientific publication process *New Engl. J. Med.* **373** 2393–5
[3] McKiernan G 2000 arXiv.org: the Los Alamos National Laboratory e-print server *Int. J. Grey Lit.* **1** 127–38
[4] Beall J 2013 Predatory publishing is just one of the consequences of gold open access *Learned Publishing* **26** 79–84
[5] Bowman J D 2014 Predatory publishing, questionable peer review, and fraudulent conferences *Am. J. Pharm. Educ.* **78** 176
[6] Butler D 2013 Investigating journals: the dark side of publishing *Nat. News* **495** 433
[7] Harnad S, Brody T, Valliares F, Carr L, Hitchcock S, Gingras Y, Oppenheim C, Stamerjohanns H and Hilf E R 2004 The access/impact problem and the green and gold roads to open access *Ser. Rev.* **30** 310–4
[8] Schultz D 2013 *Eloquent Science: A Practical Guide to Becoming a Better Writer, Speaker, and Atmospheric Scientist* (Berlin: Springer)
[9] Hawkins E 2019 Climate Lab Book https://climate-lab-book.ac.uk/ (Accessed 16 October 2019)
[10] Bornmann L and Daniel H D 2007 What do we know about the h index? *J. Am. Soc. Inf. Sci. Technol.* **58** 1381–5
[11] Ioannidis J P, Klavans R and Boyack K W 2016 Multiple citation indicators and their composite across scientific disciplines *PLoS Biol.* **14** e1002501
[12] Gopen G D and Swan J A 1990 The science of scientific writing *Am. Sci.* **78** 550–8
[13] Wallwork A 2016 *English for Writing Research Papers* (London: Springer)
[14] Stiller-Reeve M 2018 How to write a thorough peer review *Nature* https://doi.org/10.1038/d41586-018-06991-0
[15] Ross-Hellauer T 2017 What is open peer review? A systematic review *F1000Research* **6** 588

# Effective Science Communication (Second Edition)
A practical guide to surviving as a scientist
Sam Illingworth and Grant Allen

Chapter 3

Applying for funding

I require 10 000 marks.

—Otto Warburg

3.1 Introduction

Since the Renaissance Period (and until surprisingly recently), academics were often self-funded—born of a wealthy family or made rich by some good fortune or industrial endeavour. Self-imprisoned in their archetypal laboratories, many of the famous names in medieval, and even 19th century science, used their personal resources to investigate whatsoever scientific direction they saw fit. Their unfettered free thinking and growing organisation gave us much (if not most) of the fundamental understanding of the natural world and its governing physics that underpins all science and technology today. However, the proliferation of science and technology during the Industrial Revolution, and the improvements in education that came from a more progressive society, soon meant that the deep thinkers and innovators of more recent modern history needed to seek external resource to fulfil their ambitions to learn and create. And with that need for resource came the necessity to justify and define the deliverables of projects to potential benefactors, from public agencies and charities to large corporations.

Increasingly, the 'impact agenda' features strongly in any request for funding—the ends must justify the means. Long gone are the days when a scientist could simply name the price of their research endeavours as in the case of the quote at the start of this chapter by Otto Warburg to the Emergency Association of German Science in 1921. Nowadays, most academic (and plenty of other professional) roles will at some point require you to make a case to a funder or funding body to invest resource in your projects and ideas, in order to allow you the time and resources you need to pursue them. In science, this case for support (or proposal) is typically a discussion of the current state of knowledge in a specific field, which aims to highlight a new frontier or challenge that you plan to address within a well-defined and achievable project design.

A potentially successful funding application essentially needs just three things: a great idea, a great project design, and excellent communication to those that might fund it. Like everything else in this book, a successful proposal often hinges on an ability to clearly communicate a narrative to a target audience. Your research idea may well be world-changing, but if you can't convince others of that potential, you may never get the resource that you need to investigate it.

Referring to the triangle of effective communication that will be further discussed in chapter 4, writing a proposal is about self (your idea), audience (the funder) and narrative (why your idea needs investment). The advice given in this book can help you to target and communicate effectively with your potential funding agency, or other investor, and give some useful insight into how to develop scientific ideas into workable projects. But this book can't provide those exciting ideas—this spark of creative genius is still all very much down to you, which is precisely as it should be.

In this chapter, we aim to take some of the pain out of preparing a scientific proposal. From personal experience, even just the names of some of the elements of a proposal appear daunting to the uninitiated, and the process can appear very mysterious indeed. This chapter will look at the elements that comprise a modern scientific proposal and discuss how to formally develop your ideas into an achievable project, and then convince reviewers and funders of the need to invest in your work. We will begin by discussing the process of developing your ideas and your project narrative, and then discuss the funding process to offer insight into the machinations of scientific peer review and funding decisions. Much of what will be discussed is directly transferrable to any project proposal (e.g. a business case to a corporate sponsor and investors), but the focus and examples here will be on scientific project proposals presented to public funding agencies.

3.2 What makes a good idea?

This is a highly subjective question, especially as someone working in a different field may not recognise the importance or feasibility of your idea or project. Thankfully, there are typically checks and balances in the review process that mitigate for this, but this is still something to be aware of when structuring the narrative in any proposal (see section 3.4).

Under-confident researchers (especially those at the beginning of their careers) with truly great ideas can sometimes lack faith in their own creative ability, and therefore struggle to express confidence in those ideas. Likewise, there are plenty of over-confident researchers who have mediocre ideas but believe in those ideas (and their ability) so much that they can present a very strong and convincing case for support. Therefore, perhaps the most important advice we can offer is to be honestly confident. If you're not confident about your ideas and your ability to deliver on them, then it will be impossible to convey that confidence in any research proposal and it will be doomed to fail. The opposite of this would be to present a hollow front for a poor idea, something that is usually blindingly obvious in the cold light of honest academic scrutiny, though of course mistakes do happen. In other words, confidence should flow naturally from a well-reasoned idea into which you have invested the necessary thought and time.

Your goal is therefore to think of an idea (or research question), to convince yourself that it is important and has potential, and then to invest time in the practical development of that idea into a feasible, costed project. After which you will be better equipped to convey that natural and bona fide confidence to those assessing the merits of your proposal.

Fundamentally, a good idea has to represent an important new advance in some field of science, and ideally (and increasingly) provide direct and/or indirect benefit to society; for example, via public health, the economy, and/or the environment. Arriving at such an idea is typically a natural process after many years of exposure and research in a specific field such that you naturally find yourself at its cutting edge—for example after completing a PhD thesis. Identifying those ideas or scientific questions is the measure of a truly independent academic that can finally detach from the supervision of a senior academic (e.g. a PhD supervisor) and embark on their own path of investigation and discovery. True creativity and enjoyment come when we are able to push beyond our comfort zones; allow yourself the time and space to explore your ideas, discuss them with your trusted colleagues, and listen to their feedback. And then go ahead with writing your first funding proposal. It may well fail (most do, even for seasoned scientists) but don't let that stop you from trying again and again (and again). Science is built on adventurous and resilient characters that push their own boundaries, and the boundaries of knowledge in the process.

A good idea is also about the relative balance of risk and reward. Funding agencies often attempt to qualitatively (and sometime quantitatively) assess the inherent risk in any project idea, and balance this with the impact and reward that might result in project success. A truly exciting and disruptive idea with high risk of failure but high potential reward may well be prioritised over an incremental but relatively safe scientific advance. Many of the projects we receive to assess as peer reviewers are often safe incremental projects with a sound project design. However, every so often a truly exciting project is submitted that really captures the imagination, but that may need a leap of faith in order to convince that it is a valuable use of resource. Our advice here is to be careful but not to be scared. As a seasoned academic with an excellent track record (see section 3.4), reviewers and assessors may be more convinced of your ability to deliver on riskier projects, but this should not stop you from trying if you know you have an amazing idea. Just be mindful that any risky project must still have a good project design that minimises, identifies, and mitigates any risks, while also distiling the impacts and rewards that may come if the project is a success.

Let's take two examples here. One that is high-risk-high-reward, and one that is moderate-risk-moderate-reward. Another successful category might be low-risk-high-reward—but experience teaches us that such proposals are very rare indeed. The first example might be the discovery of the Higgs boson—a fundamental, force-carrying particle that, if discovered, would help to complete the Standard Model of particle physics and pave the way for all sorts of new fundamental physics and potential technological breakthroughs. At the conception of the Large Hadron Collider project, this risk was very high—there was no hint of observational evidence, the cost of experimental infrastructure and personnel was enormous, success was far from guaranteed, and the obvious pathways to new technological breakthroughs were not explicitly tangible. However, the case is (and was) compelling, attracting multi-billion Euro investment and a stream of keen researchers to the task. The second example might be something like the development of a new model parameterisation for weather prediction requiring new field data to validate it—an incremental and achievable improvement on an already good weather forecast using new understanding gained from reliably obtained field data (though colleagues at the UK Met Office may well argue that such a new advance is far from incremental….).

To summarise, have the confidence to explore your ideas, take the time to research the field of interest, and develop a practical plan (see section 3.3) that turns these ideas into a proposed investigatory project. Consider and highlight the risks and rewards and mitigate, minimise, and maximise, respectively. Get advice from trusted colleagues, and if necessary bring in the skills of those you need to help complete your project. And get thinking. Science needs you.

> **Exercise: turn an idea into a project**
>
> Write down a list of three science ideas you have that would represent novel advances in your field of interest. This could be the follow-on work you wanted to do when you finished your PhD thesis, some outstanding questions from a recent paper you wrote, or something completely different. Try to make one of these ideas something truly adventurous and risky (in terms of potential for successful investigation).
>
> Once you have this list, think about what you would need to do to investigate each of them. What equipment might you need? What data do you need? How will you obtain that data and equipment? What facilities do you need? How will you analyse the data or build your prototype? How long will each step take? What can you do yourself and do you need help from others (including people at different institutes) with specialist expertise?
>
> Put these activities into order in self-contained 'work packages' and work out the total project time and make a rough calculation of the cost. Think about what can be done in parallel, and which work packages depend on the outcomes of another.
>
> For each project, think about the risks at each step. What would it mean if you could not obtain part of the data? How could you minimise the chances of that risk becoming a reality, and what would you do if it became unavoidable? Think about the 'critical path' of your project and how you might still achieve project success if part of the project was not possible, or yielded a null result.

3.3 Finding funding

When you have decided you have a good idea and a workable plan in principle, you need to identify a suitable funder or funding agency that fits your idea (and your budget). If your idea has industrial applications, you could approach specific companies for investment and ask about their research and development strategies and how to engage with them. For public funding agencies, look up your national science funding council (or councils) and read their webpages to learn what their scientific remit is to gauge whether your project is a good fit to their advertised strategy. In Europe, there is also an over-arching European Research Council that accepts proposals from across the European Union across many different scientific themes. Often, research councils will list contact information for you to informally discuss whether your idea is eligible within the council's remit, and you will usually be pointed in the right direction if not. Finally, a good chat with more experienced colleagues will normally save you a lot of time when working out where to apply for funding; as discussed in chapter 9, mentoring is a life-long and valuable asset in any professional career.

Research councils typically release calls for proposals, or announcements of opportunity, on set advertised dates (with specific deadlines). Other calls for proposals may be open and continuous with rolling deadlines. Spend some time looking at current and past calls for proposals and announcements of opportunity to learn more about what different funding agencies are interested in, and where you might fit in to their remit. You may also be able to register for email alerts for new calls from funding agencies, and this is a good way to keep track of what opportunities are coming up without having to remember to make return visits to webpages.

Sometimes, your idea may need to be adaptive in order to respond to a relatively prescriptive call that the funder wants to commission. At other times, you may have an idea that is more suited to an open round that accepts ideas across the broader swathe of the funder's remit. But always make sure that you have identified the right place to submit your proposal to by reading the webpages associated with the funder and any call for proposals.

Perhaps the most important funding opportunity to any new academic, graduating PhD student, or early career post-doctoral researcher, are early career research fellowships. These are highly prestigious awards made to researchers with excellent ideas who show great promise for academic career development. While fellowship opportunities do also exist for researchers at various stages in their career, this chapter will focus on early career fellowships only. Such fellowships are a truly career-advancing opportunity for any aspiring new academic; they are mostly designed to fully fund your salary and your project costs at a nominated host institute for the duration of the award, and as such mean that you may be free to pursue your project without conflicting demands of teaching and other academic duties (if this is what you desire and negotiate with your host).

Early career fellowships are as much about an investment in the individual as they are about the investment in a project idea. Funding agencies and reviewers look for proposals from promising new academics with an already strong track record of research outputs, and who have an exciting idea that really capitalises on their existing skill as an independent researcher in their chosen field. Such fellowships may well be hosted at an institute where the candidate is not currently in residence, where the applicant may benefit from new skills and expertise working alongside a world-leading research group that complements their research idea. Equally, it may well be that the applicant has a good case to develop a fellowship at an institute with which they already have much experience. In either case, reviewers will look for evidence that the applicant has a strong record as an independent researcher, and that they know how to synergise with other teams and research groups to meet the objectives of an independently developed fellowship proposal.

Successful fellowship applications are often about personal independent idea development, building on a track record in a field that is well supported by the host institute. Finding a natural niche within a strong existing research group where you can truly synergise with their activity is a good formula here. In summary, if you are considering a fellowship proposal, you must also consider the role of your host group (or groups), and the mutual fit of that group to your proposed project idea. You must also demonstrate a longer-term career research strategy that justifies the investment in you as an individual academic with career potential.

Once you have identified a funding agency and a call for proposals that you intend to submit to, conduct some more in-depth research to check your eligibility, and to learn about what components to any proposal may be required in your submission. There is usually guidance and a template that you can follow to make sure that you include all the relevant information the funder needs; for example: a CV, budget documentation, and a case for support (see section 3.4). You can typically find all of this information in a funding handbook that can be downloaded or otherwise requested from the funding agency. If in any doubt, always make contact with the funding agency with any questions you have—it is part of their job to help you in this process. If you are hosted by an institute that is already eligible to apply for funding, then use the support that they can provide—find out who to talk to and let them know about your plans and learn what they can do to help you. For example, there may be a research finance office at your institute that can help you with costing your project (see section 3.5). Again, speaking with your colleagues is a good way to learn more about what support is available. You should rarely be completely alone in developing a proposal, but you may have to spend some time in finding out where this support is and how best it can be accessed.

Exercise: research your funding opportunities

For each of the ideas you explored in the previous exercise, do some research in to which agencies, companies, etc you might approach for funding, and find out how to submit a proposal to a call from that funder. Examine the remit of those funding bodies and whether that remit fits your idea. You could begin with an Internet search for 'science funding agencies', or at the homepages of your national research council.

Where possible sign up to the email alerts from any of these agencies that are of relevance and interest to your research, so that you can be kept up to date on any new announcements of opportunity. Also, take the opportunity to download any funding handbooks or guidance on preparing a proposal.

3.4 Anatomy of a research proposal

Unfortunately, there is no magic formula for what needs to be included in a research proposal. There are however, some key aspects that are essential to all proposals, and others that may be interchangeable depending on the funder. Always check your funder's handbook or other guidance to find out what information is mandatory, and if there are any templates that might need to be followed. Sometimes even formatting these documents with regards to specified page margins, font styles, and font size are mandatory and may result in outright rejection if the guidance is not followed exactly. This can be especially disappointing if you have spent several months of work developing your proposal, only to have it rejected without review due to one line in an attached CV being written in the wrong font.

Some typical components of a modern research proposal may include the following:
- Case for support—a justification of your science idea and work plan
- Justification of resources—an itemised budget linked to the work plan
- Pathway to impact—an outline of the benefits and legacy that will come from your work, and who this will affect
- Data management plan—a discussion of how you will disseminate any data to others
- Curriculum vitae—this should be written in an academic style
- Track record—an outline of how your current experience fits the proposed project
- Project management plan—a guide that provides timelines and responsibilities, and which is linked to the work plan
- Risk management plan—an identification, minimisation and mitigation of project risks, linked to the work plan

There may also be summary documents that you may be required to supply, which explain your idea to a non-specialist audience or to prepare reviewers for the detail in the rest of your proposal. There may well be other components depending on the funder and some of these elements may be merged into one document, but this will be clearly specified in the documentation associated with each specific funding call. Furthermore, while all of these elements are important, the 'case for support' is (by definition) where reviewers will look to be impressed by your project, while the 'pathway to impact' is an essential aspect of justifying why your project will benefit the wider society. There now follows a more detailed look at these two components.

3.4.1 Case for support

This is arguably the most important part of any complete proposal. It is through the case for support that you need to convince reviewers that you have an important idea supported by a workable project that is worth investing in. This can be anything from a few pages of A4 to many volumes (for very large and complex projects with multiple work packages, such as a new satellite instrument). More often than not, it is a challenge to make your case in the space allowed, which is where your ability to communicate is essential. You must efficiently, succinctly, and transparently describe your science idea to the level and range of reviewer expertise. Reviewers are usually selected by the funding agency (see section 3.6), and are typically leaders in the field in which your proposal lies. However, the subject matter of many proposals may be so cutting edge that not all reviewers will have a high level of expertise in the specific area. Therefore, you must prepare your narrative for an audience that spans from the general to the highly-specialised scientist in your field. For example, in the field of atmospheric science, a proposal about the aircraft-based measurement of greenhouse gases may be reviewed by both someone who is an expert in the mathematics of sampling theory or by someone who develops sensors to measure greenhouse gases on the ground. Clearly, each reviewer may understand some aspects of the proposal better than others, but it is your job to provide sufficient clarity such that any reviewer can both understand the general thrust of your proposal, yet also access the necessary technical detail they may be looking for in their area of expertise.

The best way to achieve this is to structure your case for support such that you 'funnel' the reader from the overall project idea and deliverables (e.g. an executive summary), through the general scientific context of the wider field surrounding your idea, before finally describe the cutting edge of that field which you plan to address and how you will do this. At each stage, it is useful to introduce and summarise each section in turn. Like any story, your case for support, and each section of it, must have a beginning (an introduction and leading summary or abstract), a middle (flowing from the general to the technical), and an end (legacy of the project). This rolling introduction and summary throughout each self-contained section really helps reviewers to digest your project ideas, and works on the following principle of triplicate repetition and memory retention: tell your audience what they are about to read, give them the information, and then summarise it for them. If you look carefully, most text books (including this one) or research papers follow this exact principle.

Reviewers are typically very busy academics who are not paid for their review work—your job is to make the narrative of your case for support easy for them to understand, as quickly as possible. A busy reviewer will likely not have the patience to read over your proposal many times to try and understand it; you must capture them on the first sentence and retain their concentration until the last. By taking them on a journey from introduction to technical detail and back to concluding summary, this will make the digestion of the information much easier for the reviewers, and will help them to commit it to memory. Ultimately, a pleasurable reading experience will put the reviewers in a better frame of mind to more objectively score your proposal; especially if you've facilitated their understanding of any technical detail by funnelling them through that detail gently and succinctly.

```
Abstract
                        Big picture, general language
Introduction

State of the art

Project details
                        Details of approach, technical language
Work plan
```

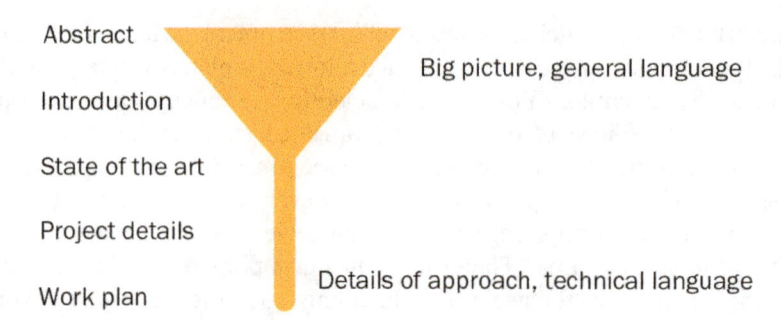

Figure 3.1. The proposal 'funnel'. This diagram outlines how your case for support can be used to create a narrative journey from the general to the specific.

A generalised structure to a case for support might resemble the illustration in figure 3.1 following:
1. A short summary of the project and why it is important.
2. An introduction or literature review that discusses the scientific context to your field (e.g. previous work), which culminates at the cutting edge that your project addresses.
3. A description of what your project will do, broken down into self-contained but cross-linked work packages.
4. A description of the deliverables and outputs of your project linked to those work packages.
5. A concluding summary on the legacy of the project and the future science that might be enabled by it.

Sometimes, project management and risk strategies may also form part of the case for support—check the funding handbook(s) of your chosen funding agency for detail.

First-time scientific proposal writers often have experience of writing peer-reviewed papers, and find it difficult to properly weight project design against scientific context, i.e. the underpinning literature review. It is important to summarise the state of play in the field that underpins your project and to identify the gap that you will fill, but this is secondary to the details of the project that you intend to carry out. Most reviewers will already be familiar with the field, therefore your introduction and literature review should be a highly succinct summary that rapidly funnels to the cutting edge that you address, before you then start describing your project in more detail. Any contextual discussion is about demonstrating to the reader that you are well informed and at the forefront of your field; providing evidence for the importance of the work you propose to do. It is not about regurgitating everything you know about your subject and exhaustively listing and discussing the merits of every paper ever published in the field. When you are limited on space, it is important to get into the project details as quickly as possible and not to bore your readers with what they may already know.

Now let's try and explain this further by way of our example of the aircraft-based measurement of greenhouse gases mentioned above, albeit in a very simplistic and abbreviated way so we can see the general flow and structure of a case for support:
1. **Summary**—Measuring greenhouse gases is important to understand climate change. Using aircraft sensors is a good way to measure them. This project will do that and provide new data to climate scientists.
2. **Context**—Lots of work has been done to measure greenhouse gases on the ground but more data is needed in places that only aircraft can get to. Complex computer models can be used to calculate emissions using this data.
3. **Work Packages**—This project will: (a) install a new instrument on a plane; (b) provide data from the new instrument in a special field campaign; (c) interpret the data using complex computer models to derive new maps of greenhouse gases and their sources.
4. **Deliverables**—This project will provide: (a) new datasets for use by scientists; (b) new understanding of greenhouse gas sources in new places; (c) papers, conference presentations etc.
5. **Legacy**—New maps of greenhouse gas sources will allow future work to update climate predictions and target policies for emissions reduction.

If you can leave the reviewer with a clear and complete story about what you want to do, why it needs to be done, and how you will do it, then you will have succeeded in writing an effective case for support.

3.4.2 Pathway to impact

A pathway to impact is where you explain how the deliverables of your project will benefit others. It is about recognising how your work fits into the wider scientific community and society, and how to make sure that you are not the only human being to know about your exciting work. It is a more detailed description of the legacy of your work, and should contain a description (pathway) of how that benefit potential will be realised in practice.

The first step here is to list the stakeholders in your research—who benefits and why. These stakeholders could be other academic beneficiaries who may wish to use your data for other projects. They could be members of various publics who will benefit directly or indirectly because of some change brought about by your research (e.g. a new drug that will improve lives, or an improvement in air quality due to new policy advice). They could be industrial partners who will use your research to develop new technology and generate economic benefit. Or they could be school-children and interested individuals who will learn about your work through targeted outreach and public engagement (see chapter 5).

The key here is to consider the wider beneficiaries of your research and then think of ways to maximise that potential through different pathways. For example, if your work has an economic benefit, how will you engage with companies who might want to use it? How will other academics learn about your new datasets? How will you inspire others? Do you need time and money to do this? Might the media be interested?

Keep in mind that your work should always be about the construction of new knowledge, and how this can be used to benefit others and the world around you. If you remember this, and take some time to think about how you can make that happen, then constructing a pathway to impact should be a natural and simple part of any proposal.

Exercise: write an impact plan

For the ideas that you developed in the first exercise in this chapter, list all of the stakeholders that you can think of for each project. These may be other research communities, specific companies, public bodies, policymakers, educators, or other publics. Write a line or two about which outcomes of your project are most relevant to each stakeholder, then think about how they might interface with your project or its outputs and how you might do this; for example via publications, conferences, workshops, stakeholder events, policy guidance, one-to-one meetings, follow-on projects, etc. How will you engage with each stakeholder and at what point in the project? Finally, write about what might come from this engagement and the benefits to you, the stakeholder, and society in general.

3.5 Budgeting

It can sometimes be difficult to know exactly what resources you will need in order to carry out a project. This is definitely an area where you should seek help from finance administrators at your host institute, who can calculate specific costs for you once you have defined what you need (e.g. two years' salary for a research assistant at Full Economic Costing).

It can often be tempting to reduce costs to make a proposal look more attractive to a funder. Much care is needed here. The best advice we can offer is to resource exactly what you calculate you will need, justify the resource, and ask for not a penny less, nor a penny more. Also, make sure that the project design fits within the budget available. Don't be tempted to propose an overly ambitious project that you cannot afford to do within the budget that is available. Reviewers and funders will look for justification of the resources requested and also seek assurances that the project is properly funded. If you have asked to conduct a field project but you have forgotten to ask for the cost of travel and accommodation, a reviewer may ask how you intend to pay for this. Equally, if you have costed a stay in a 5-star hotel and a three-course meal every evening, a reviewer may ask why this is justified.

In summary, it is more important to resource your project fully to meet all the needs of the project design, than it is to miss anything out in order to make it appear cheaper. And in all cases, your total cost must be less than or equal to the maximum budget that may be set by your funder. Your costs also need to be realistic and appropriate. It is usual to assume some contingency (e.g. inflation over the course of a project), but wild estimates of costs for items on your proposal will make you look greedy (does that laptop you need really cost as much as you suggest?). Similarly, you may be required to seek multiple quotes for expensive new instrumentation, which you may need to attach as part of your proposal. If you find that the calculated costs exceed the allowed maximum, you may need to redefine the project activities. It is far better to do this at the planning stage than to find yourself unable to carry out the project after you have received an award.

In most cases, your institute's finance administration will help you with (and usually must approve) costs for staff and other items. But the costs of any equipment, travel, or consumables must first be calculated by you and then provided to them. Furthermore, all costs must be justified against the needs of the project, and itemised for scrutiny by funders in the justification of resources information.

Finally, it can be a very useful exercise to think about and plan how you might leverage funding or contributions-in-kind from other partners or grants you may have been awarded. For example, a colleague at a different institute might offer you free access to their specialised facilities, or someone else might agree to provide you with a unique dataset and advice in using that dataset. These contributions-in-kind represent significant added value to any proposal and can help to bolster its quality; not only by adding value but by benefitting from expert support and guidance, thus minimising risk and serving to enhance impact and knowledge exchange. To this end, think carefully about how you might garner support from colleagues at other institutes, and how you might be able to work together for mutual interest and support. If such support is forthcoming, be sure to gather letters of support from those partners explaining what their contribution to your project might be, and why they are supportive of both you and your project.

3.6 The funding process

One of the most frustrating things you may have to learn to be resilient to in your scientific career, is that the majority of your proposals may simply not be funded. The reality of modern funding frameworks, both in publicly funded science and in industry, is that the pot of money is not bottomless, and there is a large pool of other excellent scientists, all with great ideas that are often in direct competition with your own. All that you can do is to present your ideas and case for support in the best way possible, rejoice and excel in success, and learn from feedback on your failures. Often, you may find that reviewers and funders may have recommended your proposal to be of high priority for funding, only to find that the finite pot of money means that other proposals higher on that priority list receive the funding available at the time.

Finding your potential funder and writing your proposal are the parts of the process over which you have complete control. Once you have submitted your proposal, the process of review and assessment begins. To give yourself the best chance possible after submission, ask senior colleagues to look over your proposal and provide an internal peer review. Do this well before any deadline and you will have time to build on their constructive comments. In any case, many institutes now require formal internal peer review prior to approval for submission to funding agencies.

The typical pathway to a successful (or unsuccessful) funding award may follow that outlined in the flow diagram shown in figure 3.2. You can find the exact pathway for your funder from their guidance information; for example, not all proposals require a face-to-face interview with a funding assessment panel, and not all funding calls allow you the opportunity to respond to reviews. Typically, your proposal will be first reviewed and scored by experts. Those reviews, sometime together with your response to them, will then be passed to a specially convened funding panel consisting of several members, who will then judge the relative merits of each proposal based on the reviews and responses and make a recommendation on priority for funding (usually assigning a score against criteria you can examine). A rank order will be defined by this panel for any single call, and the funder will then award money to those proposals down the list until the money available is exhausted. We will now briefly look at each of these stages in turn.

On submission of your proposal, a funding agency will first check for the individual's eligibility and whether the science proposed is a fit to their remit. They will then check any formatting requirements and ensure that all components of the proposal are present. Some funding agencies can be quite unforgiving at this stage and may reject a proposal without scope to put things right, so make sure you have provided everything required exactly as it is asked.

After this initial check, the funder will select and invite potential reviewers to report on and score your proposal against pre-defined criteria. You can usually find

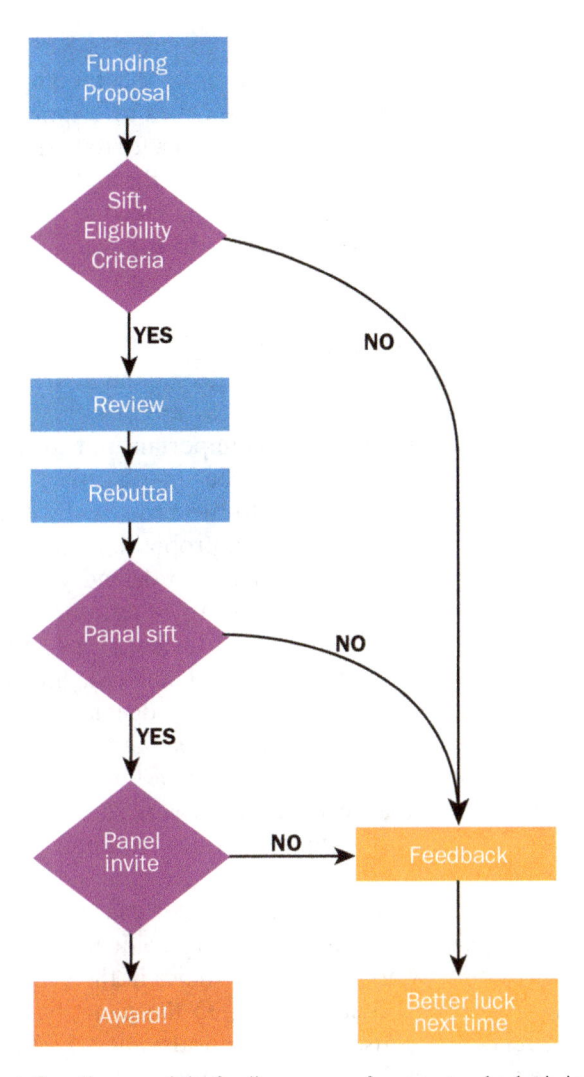

Figure 3.2. A flow diagram of the funding process, from proposal submission to award.

information on the questions reviewers will be asked and the scoring criteria in the guidance provided by the funder. It is always useful to examine these, and to make sure that your proposal clearly provides information to help the reviewer. You may also be asked to provide the contact details of potential specialist reviewers who are not connected with the project. This does not guarantee that those individuals will be approached by the funder, but it is very good practice to provide names for professionals that you may be aware of, and who may be best placed to understand and comment on your project idea. The funder can then judge the fit of any reviewers you suggest and may choose to invite them.

The pool of reviewers that a funding agency calls on is typically carefully constructed, and consists of leading experts in the fields within the remit of that agency. However, some proposals may well be reviewed by academics or professionals with only limited expertise in your area. Reviewers are usually experienced and objective and will make clear to the funding panel where the limitations in their review may be. They will be asked to comment on the scientific import of your project, your ability as a project manager to carry out the work, and the efficacy of the project design in terms of its strengths and weaknesses. They may be asked to comment on any risks, and they will be asked to fully assess the impact of the project and the justification of resources. In the case of fellowship applications, they may also be asked about the track record of the applicant and their potential as an independent researcher.

Depending on the funder and the call, you may get the chance to respond to (or rebut) any reviews to help to clarify any misunderstandings. If this is the case, you may receive any number of reviews (usually an absolute minimum of two, and typically more). As discussed in chapter 2, is important not to see your reviews as personal indictments—they are expected to be an objective discussion of your proposal and you should see your response to them as equally objective. At times, you may feel like the review is an attack on your proposal. And in many ways, that is exactly what it is supposed to be. It is your job to defend your approach (if you believe the reviewer to be incorrect) and answer any questions raised. Address each comment in turn politely, and be honest. If a reviewer has truly misunderstood some aspects of your work, the funding panel who make the final decision will read both the review and your response and reach a judgement. In many cases, the funding panel can completely disregard reviews it feels are not accurate or objective, so don't be too disheartened if you receive negative reviews. It is often better to have a detailed critical review than it is to have a very short, shining endorsement of your proposal that contains nothing that the funding panel can make a judgement on.

A funding panel is convened by the funder to prioritise and score proposals after peer review. The panel typically consists of leading academics drawn from the peer review pool whose expertise cover the range of topics in the submitted proposals, a chairperson (to manage the discussions and keep the panel on track), a secretary from the funding agency to advise on procedural issues, and a funding agency observer who may sit on several panels to monitor consistency in funding panel decisions, across the various calls administered by the funding body. Your proposal will normally be described to the entire panel verbally by a first introducer that has been asked to read your proposal, the reviews of your proposal, and your response to the reviews in advance of the panel meeting. A second introducer may then be asked to do the same in turn, following which the chairperson will then ask the introducer(s) to discuss the strengths and weaknesses of the proposal and invite others on the panel to comment. Ideally, all members of the panel will have read every proposal and are free to do so. However, in practice, each panel member may be asked to introduce many individual proposals and the total being assessed in any one panel may be very large indeed, meaning that panellists typically may only read and comment on those proposals most closely aligned with their expertise. The introducer(s) will then be asked to agree on a score that is recorded by the secretary. Having sat on several such funding panels, we can confirm that it is an exhausting, but very objective and transparent, process. The panellists will do their best to assess the relative merits of your project, and the checks and balances of the panel process and peer review help to ensure neutral objectivity. However, the process works best for you when you can really make it as easy as possible for all concerned to fully understand and access your project. Try to put yourself in the position of a panellist when writing your proposal—make sure that every word and sentence is meaningful and useful, and that its structure allows a very busy person to absorb it easily. If you don't get this right, you risk your ideas being lost in the noise.

After the panel have scored your proposal and all others, a rank order of the scores will be compiled. Many proposals may have equal scores and those that do must be placed in priority order relative to those around them. At this point, the panel may briefly re-examine those proposals and rank them again such that a rank order is achieved.

Finally, the funder will allocate funding down the list and draw a line when the money is exhausted. If your proposal is above this line (and scored fundable in principle), then well done, your project has been funded. However, the vast majority will typically be below this line. It is not unusual for fewer than one in five proposals (and often a great deal less than this) to make the funding cut. With this in mind, remember that you may have to write five good proposals before you have an odds-on chance of beating the average. Your job is to stack the odds in your favour by writing an excellent proposal based on exceptional ideas.

If you have been unlucky, you may be able to request feedback on the comments of the panel, which outline the reasons for a lack of award. Reflect on these comments, learn from them, and consider resubmitting your proposal in future by addressing any comments from the panel and reviewers. The important thing is not to give up and not to take it personally. Our role as scientists is to discover, question, and defend the truth. This is precisely what the peer review process mirrors. Don't be afraid of yourself, your ideas, or those that rightly question them.

3.7 Summary

This chapter has described how to apply for funding and how to frame scientific ideas in the form of a well-structured proposal. We have explored the machinations of the typical funding agency peer review process, and the component parts of a funding application. But most of all, this chapter has explained how any good proposal and project idea is born out of the confidence to explore your own creativity and reflect constructively on your own ideas. Be confident, be resilient, be methodical, seek advice, and don't be afraid to fail.

3.8 Further study

The further study in this chapter is related to your grant writing skills; it should make you think further about what is necessary for successful writing:
1. **Read some successful grants.** Ask the research and knowledge exchange office (or equivalent) at your institute to provide you with a selection of successful

proposals. Alternatively go to a funding body's website and look for past examples of previously successful applicants. Reading a selection of successful applications will help you to understand what is required, and will also help to develop your writing style.
2. **Make a list of upcoming deadlines.** Select a number of funding bodies that you are interested in working with and make a note of their upcoming calls and their respective deadlines. Are any of these achievable for you to apply for?
3. **Prepare a research proposal.** If you are ready, select your most promising idea and write a case for support, using the responses that you have provided to the other exercises in this chapter to help you. Note any deadlines for submission and plan your time for writing the proposal, allowing plenty of time for development and discussion with trusted colleagues.

3.9 Suggested reading

Scientific Writing and Communication: Papers, Proposals, and Presentations [1] covers both writing and oral presentation in significant depth. As one of only a few dedicated books that cover proposal writing, this resource is highly relevant to this chapter. *Writing Successful Science Proposals* [2] also provides very useful advice for how to get your ideas funded, while *Getting Funded: The Complete Guide to Writing Grant Proposals* [3] focusses entirely on proposal development from scoping an idea to submission and review of a full proposal; it also deals with budgeting and human resourcing. Finally, *Proposal Writing: Effective Grantsmanship* [4] whilst written by academics, covers proposal writing for submission to a wide range of potential funders other than national scientific funding councils, including corporate sponsors and philanthropic foundations. It provides some useful guidance on choosing an appropriate funding agency and also deals with more logistical aspects such as budgeting.

References

[1] Hofmann A H 2014 *Scientific Writing and Communication: Papers, Proposals, and Presentations* (New York: Oxford University Press)
[2] Friedland A J, Folt C L and Mercer J L 2018 *Writing Successful Science Proposals* (New Haven, CT: Yale University Press)
[3] Hall M S and Howlett S 2003 *Getting Funded: The Complete Guide to Writing Grant Proposals* (Portland, OR: Continuing Education Press)
[4] Coley S M and Scheinberg C A 2008 *Proposal Writing: Effective Grantsmanship* (Thousand Oaks, CA: Sage)

IOP Publishing

Effective Science Communication (Second Edition)
A practical guide to surviving as a scientist

Sam Illingworth and Grant Allen

Chapter 4

Presenting

Nothing in life is to be feared, it is only to be understood. Now is the time to understand more, so that we may fear less.

—Marie Skłodowska Curie

4.1 Introduction

Sooner or later in your academic career you are going to be asked to give a scientific presentation. This may be at a weekly research group meeting, or at an international conference; each of which have their own unique set of challenges to address. Presenting to any audience, and in any context, can be a difficult and demanding exercise, and despite what some people might say, it requires practice. Colleagues that profess to making it up as they go along, or 'winging it', are either seasoned professionals who have delivered similar material many times before, or else they are deluded and are likely not as good at presenting as they may think.

There are various audiences that scientists need to communicate with, and we need choose how to present our narrative accordingly. We must also consider how we present ourselves in the process. Communicating to other audiences is addressed elsewhere in this book, mainly in chapters 5 (non-scientists), 6 (the media), and 7 (online). The focus of this chapter is on that of presenting your work in person to a scientific audience, centred around practical advice designed to help you become a more effective public speaker. The best piece of advice that we can give is perhaps the most obvious: practice makes perfect. As the noted American author Mark Twain once said, 'It usually takes me more than three weeks to prepare a good impromptu speech.' Being an effective presenter is not a 'dark art', nor is it a gift that

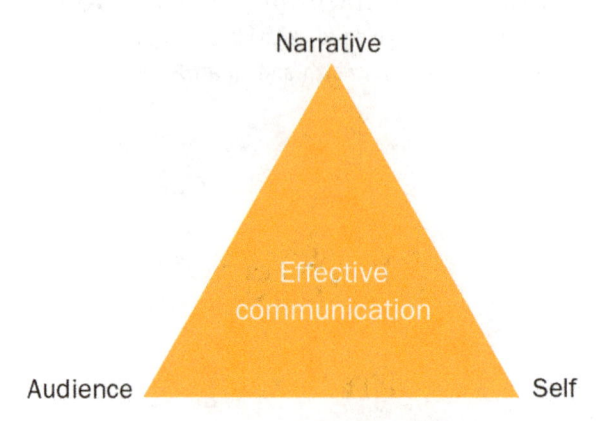

Figure 4.1. The triangle of effective communication. This triangle highlights the three key aspects that need to be considered in order to communicate effectively: the 'narrative' (what you are saying), the 'audience' (who you are saying it to), and the 'self' (how you are saying it).

certain people are innately born with; rather it comes from taking the time to learn and develop the skills and approaches that are discussed in this chapter.

4.2 A three-way approach

There are many theories on the best way to create and deliver a presentation, but the approach discussed here is based on the work of Edward Peck and Helen Dickinson [1], who outline three key properties that need to be considered in order for effective communication to take place: the narrative, the audience, and the self.

Figure 4.1 outlines this three-way approach in the form of a geometric shape: the triangle of effective communication; without all three of its components, the concept at the centre cannot exist. To communicate effectively, you need to consider the 'narrative', the 'audience' and the 'self', and if just one of these vertices is missing then the effectiveness of your message will be reduced. We will now discuss in turn how best to consider these three key properties.

4.2.1 Developing your narrative

The ultimate purpose of a scientific presentation is to communicate a message to your intended audience. This message might be that you have found some interesting findings as a result of your research, or that you have an update on some recent challenges that you have been experiencing. In order to ensure that your audience have taken away from your presentation the message(s) that you intended, first you need to do a little planning. A very effective method of doing this is outlined in the following exercise.

> **Exercise: three take-home messages**
>
> Think about the next presentation that you are going to give. Now, imagine an idealised world in which your audience leaves the presentation knowing exactly what you wanted them to know. What would that be? Write down the three key, take-home messages for your presentation. For example, if you were giving a presentation entitled 'Effective Communication in Presentations', you might want your audience to leave knowing about the importance of the narrative, the audience, and the self.
>
> Once you have written down these three take-home messages, use them to create the structure for your presentation. We also recommend including these take-home points on the final slide of your presentation (see section 4.5), as doing so will serve to further highlight their importance to your audience.

Determining the take-home messages and then using this to structure your narrative shouldn't be restricted to presentations. This technique can also be utilised in any form of communication, from international teleconferences to one-to-one meetings with a supervisor or line manager. Next time you have something that you need to communicate, first take the time to work out what your key take-home messages are; it will help you to align your arguments and it will make it far more likely that you achieve what you set out to accomplish.

Once you have your take-home messages it is time to start building the narrative that will allow for them to be communicated in a succinct and logical fashion. Giving a presentation is effectively like telling a story, which means that the same basic concepts of narrative that are applicable to storytelling can also be applied when structuring a presentation.

In their most basic format, stories have a beginning, a middle and an end, and in this respect scientific presentations are no different. In a scientific presentation, this is often about setting the scene and telling your audience what you are about to tell them (the beginning), telling them something in detail (the middle), and then telling them what you just said (the end). As discussed in chapter 3, this principle of triplicate repetition and memory retention helps people to retain and understand information; especially complex technical information that is typical of many scientific presentations.

You should begin by introducing your audience to the narrative, i.e. by providing a summary of the background and context to the scientific field your work relates to. Without this introduction you run the risk of alienating your audience, but similarly if this introduction is overly long then you are in danger of losing their attention. The introduction is also necessary to justify why the story that you are about to tell is worth listening to. An effective way to link your introduction to the next part of the presentation is to pose a question (or hypothesis) that you plan to answer later; doing so prepares the audience mentally for what they will hear next.

The middle part of a story is where the crux of the plot takes place and develops. Having laid the scene with the introduction, the storyteller is now able to explore the more interesting elements of the narrative. In a scientific presentation, this is where the methods, results, and analysis would be found; having explained the context and justification in the introduction you now have the opportunity to take the audience with you on your journey of discovery. What did you find and how did you find it? What do you think that this might mean? Were there any unexpected or surprising results?

The end of a story is where the storyteller skilfully gathers together all of the different elements into a final passage which gives insightful context and objective interpretation to the preceding narrative. In a scientific presentation it is the conclusion that fills this requirement. Having laid out the experimental process and your reasons for doing so, did you succeed in answering your original hypothesis, how confident are you in your interpretation and conclusions, and what might those conclusions mean for any future research?

An overview of a scientific presentation as a linear story is shown in figure 4.2. There are of course some narratives that make a conscious decision to leave things relatively open-ended (see *Gravity's Rainbow* by Thomas Pynchon or *Everything is Illuminated* by Jonathan Safran Foer for literary examples). The reality of scientific research means that in many instances there is no 'neat ending' to these stories either, but rather a series of questions that await further analysis. The skill in telling your story is to leave your audience wanting to find out more, while prompting them into offering useful suggestions. This particular linear storytelling device is also dependent upon the audience. For example, when giving a presentation to journalists, it may be more useful to begin with the take-home messages; this will be discussed further in chapter 6.

4.2.2 Understanding your audience

The second vertex of the triangle of effective communication is your audience. Without an audience you may as well be giving a presentation to a brick wall, and sadly this still appears to be the favoured approach for a significant number of scientists. Talking about your research is an excellent opportunity to promote

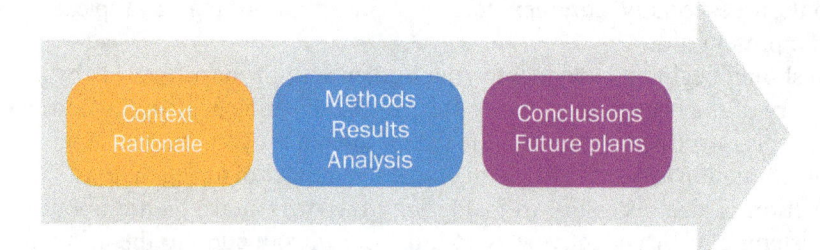

Figure 4.2. Telling your science story using a linear narrative for a scientific audience. Using this approach, your presentation should have a clearly defined beginning (context and rationale), middle (methods, results, and analysis), and end (conclusions and future plans).

yourself, and to impress the importance of your work upon the people in the room. However, if you are speaking as if no one is there, then that is likely to be the number of people interested in what you have to say.

There appears to be an unwritten understanding that speaking in front of your peers is a terrifying experience. While we will deal with nerves in section 4.3, it is worth taking a moment to reflect on this misconception. In almost all instances, the audiences that you will be presenting to are understanding, considerate, and willing you to do well. There are of course exceptions, but you are not the opening act for a provincial stand-up comedy tour. The audience is not baying for your blood, they are almost certainly there because they want to hear what you have to say, and most of them will be sensitive to the fact that they have (probably very recently) been stood in your shoes.

Given the understanding, considerate, and patient nature of your audience, it would surely be unfair to dismiss them entirely out of hand and to speak as if they were simply not there. Your audience are an integral part of your presentation, and by working with them, rather than ignoring them, you stand a much better chance of not only effectively communicating your take-home messages, but of also enjoying yourself in the process.

Now that you have decided to at least acknowledge your audience, you need to consider their needs. One of the most effective ways to do this is to speak in a language that they understand, and the following exercise has been designed to help you achieve this.

> **Exercise: know your audience**
>
> Imagine that you have one sentence to explain your research to a five-year-old child. You should avoid using any scientific jargon, or indeed any words that would not be understood by this target audience. Once you have done this, read it out loud and see if it would actually be understood by a five-year-old, or even better find an actual five-year-old and see if they understand what you are talking about. If not try again.
>
> Next, write one sentence that explains the crux of your research to a non-scientific, but adult audience. Once again, avoid using any scientific jargon, or indeed any words that would not be understood by the target audience. To make this easier, try using the 'up-goer five text editor' [2], which restricts you to using the one thousand most used words in the English language (this tool is based on a comic from xkcd.com [3], which presents a diagram of the Saturn V rocket using only the one thousand most used words in the English language). Once you have done this, test out your sentence on an adult non-scientist and see if they understand it. If not try again.
>
> Continue this exercise by writing two sentences: one that summarises your research to a scientist who works in a different field to you, and one that explains it to a scientist who works in the same field as you. Try these sentences out on their respective audiences, and then use them to help you to form the take-home messages of a presentation (depending on your audience), as outlined in the previous exercise.

Considering the needs of the audience that you will be presenting to (and the languages that they speak), will help you to avoid either patronising or baffling them. Obviously, with larger audiences, it might be difficult to cater to the needs of every individual. However, by considering the scientific expertise for the majority of the people in the room, you will help to ensure that you maintain their attention throughout.

Considering your audience does not only mean speaking in a language that they can understand, but also involves highlighting the most relevant parts of your research. For example, it would likely be far more interesting to an audience of environmental scientists if you were to tell them that your research involved 'studying the chemistry of water that drips inside caves to find out what the climate was like in the past', than it would be if you simply told them that you 'use instruments to study the chemistry of water that drips inside caves'. Likewise, give consideration to what aspects of your research your audience might need to know more about, or which sections require less attention. For example, if you were presenting your research on climate change and bird migration to experts on the subject, they may require less of an introduction to the topic than if you were speaking to climate change scientists with no knowledge of bird habitats or behaviours.

This consideration of your audience's experiences and attitudes might also be referred to as 'framing'. In essence, framing theory suggests that how something is presented to the audience (i.e. the frame) influences how it is processed [4]. Framing involves explaining and describing the context of the problem to gain the most support from your audience, and so understanding the needs and experiences of your audience is key. A relatively well-known example of the framing effect is a 2009 study which found that while only 67% of PhD students registered early for a particular conference when doing so was presented as a discount, 93% did so when the emphasis was instead on a penalty fee for late registration [5].

A final issue to consider with respect to your audience is the use of jargon in a scientific presentation. Introducing a large collection of words that are alien to your audience is the quickest way to disengage them with your subject matter. There is nothing wrong with introducing new terminology, but be sure to explain exactly what you mean by it, ideally by contextualising the term. Of course, jargon isn't just those words that are not understood, but are also words or phrases that may have a different meaning depending on the field of expertise. For example, if you were talking about long timescales, the understanding of this term would vary considerably between a meteorologist and a palaeontologist.

4.2.3 Managing yourself

The final vertex of the triangle of effective communication is you. Without you there is no presentation, and as such you should consider how to present yourself when addressing your audience with your narrative.

At first, the concept of managing, or presenting, yourself can seem quite abstract. However, it can be easily broken down into a handy acronym, as shown in figure 4.3.

If you are able to master all four of these components, then you will have succeeded in managing yourself. Taking each one in turn:

Stance. The way in which you physically position yourself is vital in positively reinforcing how confident you appear as a presenter. Your stance also concerns any nervous tics or involuntary movements; for example, do you have a tendency to jangle the keys in your pocket when you are talking, or to pace up and down the room looking at the floor? These movements can be very distracting to your audience, and they can quickly become the focus of attention, at the expense of your carefully crafted narrative. Any movement that you do make should be deliberate and purposeful.

Stance

Assurance

Voice

Eye contact

Figure 4.3. The four components that are needed to successfully manage yourself when giving a presentation: stance, assurance, voice, and eye contact.

If you are someone that naturally uses a lot of gesticulations, then don't be afraid to use them as you talk. If you don't, then it will probably feel unnatural to you, which will affect your confidence and ultimately lead to a poorer presentation. Similarly, if you are someone that does not normally use gesticulations, then avoid using them when you present, as otherwise they will become an unwanted distraction to both you and your audience.

Standing might not always be possible or preferential; for example, if you are in a small room in an informal setting then it might be more appropriate to talk to your audience from a seated position. In all instances you should try and determine the layout of the room in advance of your presentation, as this will help you to feel more confident and will also ensure that you can better account for the needs of the audience. On visiting the space ask yourself questions such as: is there an ideal position which allows most of the room to see you? Is there a lectern to make use of? Is there a fixed or movable projection screen or monitor?

Assurance. Being confident should not be confused with being arrogant. Your demeanour should highlight the fact that you are authoritative in your subject matter, but also that you are approachable. Despite what you may have heard there is absolutely nothing wrong with injecting a bit of your personality into a presentation, in fact many audiences will thank you for it. However, consider both your audience and yourself in this process. Don't do anything that doesn't feel natural to you, or which would leave you feeling awkward, as this will be picked up on by the audience.

Being an assured presenter does not mean that you have to be an extrovert, it is about carrying out the necessary preparations so that when you speak you are in a place that feels comfortable to you. For example, some presenters use humour to great effect, whereas others should leave it well alone. Above all else, be yourself, and have the confidence to stand in your own truth. So, put aside that nagging voice in the back of your head, and remind yourself that you *are* an expert in your field, that you *do* have something worthwhile to say, and that you *will* present it in an accessible and engaging manner.

Voice. The voice is the most versatile and vibrant tool at a presenter's disposal. Take time in getting to know it. Consider how your pitch, rhythm, tone, and volume can affect your delivery. You should also look after your voice, which means warm-up exercises, and avoiding alcohol, nicotine, and caffeine immediately before giving a presentation, as all of them will put a strain on your vocal chords. If you have a naturally quiet voice, then don't worry, as that is what microphones are for. If you have a booming voice and think that you don't require a microphone, you should still make use of one if it is available. The reasons for this are two-fold: firstly, the acoustics of the room mean that you might not be heard very well from every location, and secondly people might be making use of a hearing loop, which is directly connected to the microphone.

Eye contact. Try and make eye contact with everyone in the room at least once during your presentation. However, be careful not to be too intense, or to focus on any one person for a prolonged length of time, as this can lead to rather awkward situations. Instead, try to scan the room by engaging with people in turn. Doing so will help people to stay alert and engaged, allowing you to convey a sense of interest in your audience. The use of eye contact can also be a very effective way of dealing with nerves (see section 4.3).

4.3 Dealing with nerves

One of the most intimidating challenges that needs to be overcome when giving a presentation is nerves; specifically, how to deal with stage fright. Well-worn advice on this topic is that everybody suffers from nerves, and that you simply need to harness this nervous energy in order to give an effective presentation. While elements of this may be true, it is not particularly practical, nor is it helpful to the many would-be presenters whom may be perfectly capable of rationalising their fears, but are still unable to master them.

The best way to prepare for a presentation is to practise it until you are completely comfortable with both the content and the delivery. Often, nervousness arises from the fear of the unexpected. If you are confident with what you have to say and how you intend to say it, then you will be far less likely to get nervous.

It is advisable to practise your presentation until you are able to deliver it without notes, as doing so will greatly help both your nervousness and your engagement with the audience. Furthermore, if you don't have any notes then you won't have anywhere from which to physically lose your place. Similarly, it is recommended to avoid simply learning a script, as doing so is likely to affect your delivery and could lead to further nervousness if you lose your train of thought, or accidentally stray from what you have memorised.

A tried and tested way to learn your presentation is to perform it about five to ten times, avoiding notes and scripts, and aiming to include a few key points and phrases that you can iterate with every repetition. This will result in the most natural delivery style, and will also present you with the least stressful situation in terms of recollection during the actual delivery process. If you are using slides, then practising without them is also a good idea, as it will allow you to imagine where the transitions are without an over reliance on looking at the screen.

If speaking in front of a small group of people does not faze you, but the idea of talking to a larger audience fills you with dread, then practise in front of a gradually increasing audience size. Start off by presenting to a small group that you feel comfortable with, and then increase this until you are speaking to twenty to thirty people in a room; for example, during a weekly research group meeting. At this point it might be difficult to continue to practise in front of an incrementally increasing audience size, but the confidence that you have gradually built up should help you to make the leap to a larger number of people. Another tactic for dealing with large audiences is to make a note of where a friend or colleague is sitting. Begin by making eye contact with this person and directing your presentation to them, then gradually start to expand your gaze out to the rest of the audience as you gain confidence, returning back to your friend if your nerves return.

4.4 Rhetoric

The Ancient Greek philosopher Aristotle defined rhetoric as 'the faculty of observing in any given case the available means of persuasion' [6]. Rhetoric was originally abhorred in Ancient Greece, on the grounds of it being used for the promotion of the subjective truth of the speaker, rather than the objective truth of the argument (as is the case in dialectics). However, Aristotle saw the necessity for rhetoric, realising that those who were arguing for the search of an absolute truth could do so using the same tools as those who were pedalling their own agendas. While today rhetoric may be seen by many to be a euphemism for 'all style and no substance', or as the exclusive tool of 'sneaky politicians' and 'scurrilous journalists', this need not be the case. It is useful to understand the three basic elements of rhetoric, as doing so will help you to develop and deliver more effective presentations.

Ethos is an appeal to **ethics**, and it is a means of convincing an audience of the character or credibility of the speaker.

For example, when applying for a job you would want to convince the interviewer of your suitability for the role, and might draw on previous experiences and responsibilities to convince them that you have the necessary expertise.

As scientists, we often forget that we are world-leading experts in our field, and as such that we are credible sources that are worth listening to when discussing our work and research. As discussed earlier in this chapter, you should be confident in your expertise when discussing your field of scientific research. Introducing yourself at the beginning of your talk as person XXX from research institute YYY is an excellent way of reminding both yourself and your audience of your credibility, and represents a perfectly practical way to begin a presentation, especially to an audience who might be sitting through many consecutive presentations (such as at a conference or symposium).

Logos is an appeal to **logic**, and is a way of convincing an audience by reason.

For example, when explaining to an audience the validity of your data, you might take them through a step-by-step account of your methods, in which you remove all doubt that what you have observed is spurious or unrepeatable.

As scientists, we construct our research using a tried and tested method. We establish a hypothesis, test that hypothesis, and then accept or re-visit our hypothesis based on the results of our tests. This is a logical process, and providing that we conduct our work in an ethical manner (see chapter 9), and present our research accordingly, then we have this aspect of rhetoric covered. Discussions that are devoid of logic are baseless, and it is such arguments that tend to give rhetoric a bad reputation. As scientists, if we see claims by politicians, journalists, or even other scientists, that are not backed up by rigorous arguments, and independently verified results (where appropriate), then we have a responsibility to challenge these claims. Doing so helps to protect the more vulnerable members of our society from arguments that are incorrect and potentially harmful.

Pathos is an appeal to **passion**, and is a way of convincing an audience of an argument by creating an emotional response.

For example, when having an argument with a loved one you might choose to recall a particularly hurtful example of when you had been let down by them in the past; pathos is an appeal to all emotions, not just positive ones.

As scientists, this is the aspect of rhetoric that we tend to be weakest in addressing. Many of us are told that when we present our scientific findings we must only present 'cold, hard facts', but by taking such a dispassionate approach to our research we are arguably alienating ourselves from others. While we should remain fully objective when we are conducting our experiments and testing our hypotheses, when we are talking about the implications of our findings then we should stop thinking that we have to behave like robots; it is ok to get angry, happy, elated, and even upset when we talk about our research. For example, scientists who research global warming and climate change are still affected by the negative impacts of anthropogenic global warming and climate change. When talking about our research, we should express how it makes us feel; doing so helps to humanise our profession and to bridge the gap between 'science' and 'society'.

While the use of rhetoric most readily lends itself to the construction of your narrative, there are still many occasions in your interaction with the audience and the consideration of yourself to which it can be applied. For example, in the UK many dignitaries and celebrities wear a poppy when they are being interviewed on TV around the time of Remembrance Sunday. This indicates to the audience that this person is respectful (ethos), while also creating an emotional response in the viewer (pathos), depending on the connotations that they associate with the poppy in this context.

4.5 PowerPoint

When being subjected to a particularly laborious or poorly structured scientific presentation, an audience might complain about 'death by PowerPoint'. However, such statements are unfair, as it is not a piece of software's fault that the presenter is ineffective. The only thing that PowerPoint, and many similar pieces of software, may be guilty of is being an extremely versatile and easy-to-use toolkit. Here are some tips that will help to ensure that you get the most from this presentational scapegoat:

1. **PowerPoint is not a substitute for you**. At most it should be thought of as a visual aide-mémoire. It is a piece of software, not a sentient being. Ultimately it is you that will have to give the presentation, not PowerPoint.
2. **When it comes to text, less is more.** Frankly, text-only slides are boring. They are not only visually disappointing, but they also distract the listener from what you are saying while they attempt to simultaneously read the contents of a slide. Text should be kept at a minimum in any one slide, in favour of a verbal narrative that is supported by suitable graphs, diagrams, and images.
3. **Avoid PowerPoint Karaoke.** If you do have to use text, please avoid reading it verbatim from the screen. Your audience is perfectly capable of reading text for themselves. Reading text from a slide is very rarely a good idea, but perhaps there is one exception: as a means to place verbal emphasis on a take-home message.
4. **Use your take-home messages for the final slide.** Once you have determined what the take-home messages are for your presentation (see section 4.2.1), use these as a single bullet-pointed list and leave it up as the final slide in your presentation; your audience will be grateful for providing them with a helpful summary that they can easily take note of. Your final slide should not be one that asks for questions (the audience are unlikely to need reminding of this fact), or which thanks your collaborators (this can be done on an earlier slide).
5. **Check your spelling and grammar.** PowerPoint presentations with typographical errors will reduce your credibility as a speaker. Take the time to make sure that there are no errors in your slides, and if it is possible then get someone else to check them through for you.
6. **Think about slide design.** Choose your theme carefully. See section 4.8 for further advice on colour schemes, but pick a font design that can be easily read from a distance. Likewise, it is advisable to use a single, easy-to-read font throughout your presentation. Ideally all of your slides should have a similar layout, and where possible should also include slide numbers, as this makes it easier for the audience to make a note of anything that they might want to discuss with you later.
7. **Consider your aspect.** When designing your slides, consider the proportion of the slides. Try to find out in advance what the aspect ratio is of the projector that you will be using (normally 4:3 or 16:9), and design your presentation accordingly. If you are unsure then make two versions of your presentation; it might take a little more time but failing to do so will negatively affect your images and layout.

8. **Brand your slides**. Include your institutional logo at the top right, and your Twitter handle (or email address) at the bottom left of each slide. Doing so will allow people to find you online and/or contact you directly after your presentation.
9. **Embed any audio and video files into the slides.** This means that you do not have to worry about saving files in certain folders, and also avoids the need for a connection to the Internet.
10. **Make sure everything works.** You should always take the time to make sure that your presentation works correctly, including the streaming of any videos and/or audio. Arrive at the venue early and get everything set up, as this will allow you to avoid any unnecessary technical distractions. Similarly, make sure that none of your images are pixelated, and also that you have any cables or adaptors that you might need to connect your laptop to the projector.

There are many other different forms of presentation software that are available to you aside from PowerPoint. It is therefore recommended that you try out a number of different pieces of software until you find one that you feel most comfortable with, and which affords you the most assurance in your delivery. Listed below are three examples that are well worth investigating further:

1. **Prezi** [7]. Unlike PowerPoint, Prezi is not constrained to rectangular slides. Instead it focuses on the construction of frames of different shapes and sizes, which can be zoomed in and out of to create an aesthetically pleasing visual. A word of warning though: be very careful not to overuse the zoom function (the same can be said for animations in PowerPoint), as this can have the unintentional effect of making your audience feel quite nauseous.
2. **Kahoot!** [8]. An interactive piece of software that allows audience members to vote or answer questions. It is free to use, and does not require the audience to own any fancy pieces of kit other than their smartphone, which they use to input their responses. These responses can then be downloaded for later analysis and evaluation. If using Kahoot! you will need to check that there is both a strong Internet connection for the presenter, and a strong mobile broadband signal for the audience.
3. **Poll Everywhere** [9]. Similar to Kahoot! in that it encourages audience members to interact with the presentation using their mobile phones. Unlike Kahoot!, it also supports the use of text messaging, so audience members without smartphones can also get involved. There is a free version of the software, but for larger audiences a paid for pro license is necessary. Again, a good Internet connection is essential for the presenter.

Finally, you could consider using no slides at all, instead focussing on delivering your message to the audience in an effective and engaging style. This may not be suitable for technical presentations requiring graphical data, but it can be a far more engaging method of delivery for less technical presentations. You might alternatively consider the use of a single image summarising your key, take-home messages.

Whatever piece of presentational software you decide upon, find one that you feel comfortable with, and which will help you to reinforce your narrative, rather than one which will cause the audience to become distracted.

4.6 Timings

You will often be in a situation where you have a limited time in which to deliver your presentation. If this is the case, then it is very important to stick to your allotted timeslot. It is extremely impolite to overrun, as it gives the impression that what you have to say is more valuable than what others have to contribute. It can also impact on the general running of an event (e.g. in conferences with parallel sessions), and you will make no friends for delaying anyone's lunch or coffee break. In conference situations, it will usually fall upon an allocated chairperson (or chair) to make sure that the session is running to time, but you should not rely upon them to drag you from the stage once your time is up. Instead, practise your presentation with the timing firmly in mind. And don't overstay your welcome.

In order to perfect the timing of your presentation, it is advisable that you first practise it with a stopwatch clearly visible (PowerPoint provides such a tool). As you gain confidence in your presentation and timings, gradually start to stop looking at the stopwatch, until you no longer need it. When you come to give the presentation itself, the chair will normally offer you a 2 min warning (or longer depending on the length of the presentation). If they do not, be sure to ask for one, as it can help you to focus. Many presentational facilities will also provide some form of stopwatch that you can refer to throughout your talk, if required. When you come to present, adrenaline has a tendency to make you speak faster than you do when rehearsing. However, a faster pace should only be considered and not relied upon, i.e. do not prepare a 20 min presentation for a 15 min timeslot in the hope that nerves will bridge the gap.

4.7 Answering questions

One of the most nervous elements of giving a presentation is the knowledge that you are probably going to have to answer questions about what you have just said. In almost all circumstances this will be via a formal Questions and Answers (Q&A) session at the end of your talk, although there are instances (normally in more informal environments, such as group meetings) in which you might be asked questions during the presentation itself. The following advice is still mainly relevant in these situations, but take special care not to let a particularly difficult question upset the flow of your presentation:

1. Where possible take three questions at once. This is a trick that is commonly employed by politicians. Doing so allows you to answer the easiest question first, and gives you time to ruminate over the more difficult ones.
2. If you don't feel confident enough to answer any question in such a formalised setting, then offer to speak to the questioner 'off-line'. This simply means that you can speak to them in a one-to-one environment

(e.g. over coffee at a conference) giving you more time to think, and thus resulting in more useful discussions.
3. If you don't feel entirely comfortable with a question, but still want to address it, then try subverting the question slightly. This will allow you to focus on something that you are more comfortable with, but should still give the questioner the satisfaction of being answered.
4. If you do not know the answer to a question, then do not be afraid to say so. There is no harm in not knowing everything, and it may be that the questioner has considered an angle that you had not previously imagined. Offer to speak to them off-line to further probe their line of enquiry; it might be that they are able to significantly help you with your research. It might also be that they have simply misunderstood what it was that you were saying.
5. Prepare some additional material. If you know that your presentation is likely to raise questions that you are unable to address in your allotted timeslot, make sure that you have some material (be it in the form of additional slides, handouts, or simply a well-rehearsed riposte) prepared that will allow you to sufficiently address them.
6. Prepare for questions that you think you might get asked. By practising your presentation, you will become aware of which areas might require further explanation, or else which topics are potentially the most contentious. If you have the time, practise in front of some colleagues and see if you can address their questions, as these are likely to be similar to the ones you will encounter when you give your presentation 'for real'.

Asking other people questions after their presentations is also an important skill that takes practice and consideration. Here are a list of **DO**s and **DON'T**s that you should consider before asking a question:

DO make sure to preface your question with a compliment regarding the nature of the talk or the effectiveness of the speaker. You would expect the same.

DO take notes of the presentation, and use them to make a list of suitable questions to ask at the end. This is especially important if you are the chair of a session, where you have a responsibility to ensure that at least one question is asked.

DO consider talking to the presenter after their talk instead of during the Q&A session, especially if you have a difficult question, or multiple questions that require further discussion.

DON'T ask a question simply to let everyone else in the room know that you have been listening, or that you have expertise in this field.

DON'T use a question to advertise your own work.

DON'T ask any question that you cannot formulate succinctly.

4.8 Poster design

Presenting a poster is often seen as secondary to giving an oral presentation, especially at major scientific conferences. However, this is simply not true, as presenting a poster generally provides you with the opportunity to talk in detail about your research, potentially in a more informal and relaxed environment. As with giving an oral presentation there are some general **DO**s and **DON'T**s that you should follow in order to make sure that your poster stands out, and that you maximise the use of your time:

DO use your poster's structure to tell a story. Consider the layout so that it presents the reader with a logical flow, from rationale and results to analysis and conclusion.

DON'T have too much text. As in any oral presentation, your poster should mainly consist of images and diagrams highlighting the work that you have done, with a few lines of explanatory text. In most poster sessions you will have the opportunity to discuss the finer details of your research in person, and you can always prepare some handouts with further information for the reader.

DO stand by your poster for the allotted time period. Even if there is no initial interest, this might be because you are at the end of a row, or because a parallel session may be overrunning.

DON'T just follow the same old template design of everyone else in your research group. Consider using pictures from your research, and adding an element of personalisation. Also include any insignia that is required; for example, logos from funding bodies or research institutes.

DO bring some business cards and printed handouts of your poster with you. Business cards can provide people with an easy way to contact you after the event. You can also leave handouts or printed QR codes (see below) with your poster during the times that you are not required to stand with it, or for when you have to take a comfort break.

DON'T use a colour scheme that makes the poster difficult to read. Use a colour wheel (see figure 4.4) to help you match up complementary colours (i.e. those that are opposite each other on the wheel). For example, if you are using a predominantly dark green background, then red text will be the easiest to read, and vice versa. Also, be careful not to overuse colour as it can easily become distracting.

DO consider including a QR code on your poster which links to a webpage that contains more information for the reader (e.g. your personal website or a recent publication). QR codes can be made for free and in a matter of seconds using online software [10].

DO give proper consideration to the use of font. Certain fonts such as Comic Sans should be avoided altogether, as they can be very difficult to read.

DON'T pressure people the second that they arrive. Give them the chance to read your poster, and let them know that you are there to answer any questions that they might have; you are a scientist, not a salesclerk on commission in a boutique jewellery store.

DO take the time to get to know your audience. In oral presentations you normally have to make some general assumptions about the level of scientific understanding that you audience possesses. In a poster session you can instead ask your audience about their expertise and experience. This will help to eliminate embarrassing moments in which you explain the first principles of your research to an expert in the field.

DON'T be one of those people who prints out their poster on several pieces of A4 and then fastens them together with sticky tape. Find out what the dimensions of the

Figure 4.4. An example or a colour wheel. Complementary colours are those colours which are opposite one another on this wheel, and which can be used to provide the best contrast for reading coloured text on coloured backgrounds. This image has been obtained by the author from the Wikipedia website (https://commons.wikimedia.org/wiki/File:BYR_color_wheel.svg), where it was made available by Sakurambo under a CC BY-SA 3.0 licence. It is included within this article on that basis. It is attributed to Sakurambo.

poster board are in advance and design your poster accordingly, factoring in the options for orientation (i.e. portrait or landscape).

> **Exercise: you be the judge**
>
> Three posters are shown from figures 4.5–4.7. Which of these posters do you think most closely adheres to the rules for good poster design that are laid out above, and which of them needs the most attention? What would you improve about these posters, and which do you think illustrate good practice in both design and storytelling?

Some conferences offer a combination of oral and poster presentations, in which you will normally be given one or two minutes to verbally advertise your poster before the session begins. These short pitches should not be treated as a challenge in which you try to cram a 15 min presentation into 60 s. Rather, they should be seen as an opportunity to entice your audience and leave them wanting to find out more in the ensuing poster session.

Some poster sessions now also adopt a digital format, in which participants are invited to present their work using interactive touch-sensitive screens. As with learning to use the right tools for your oral presentations, learn how to use the

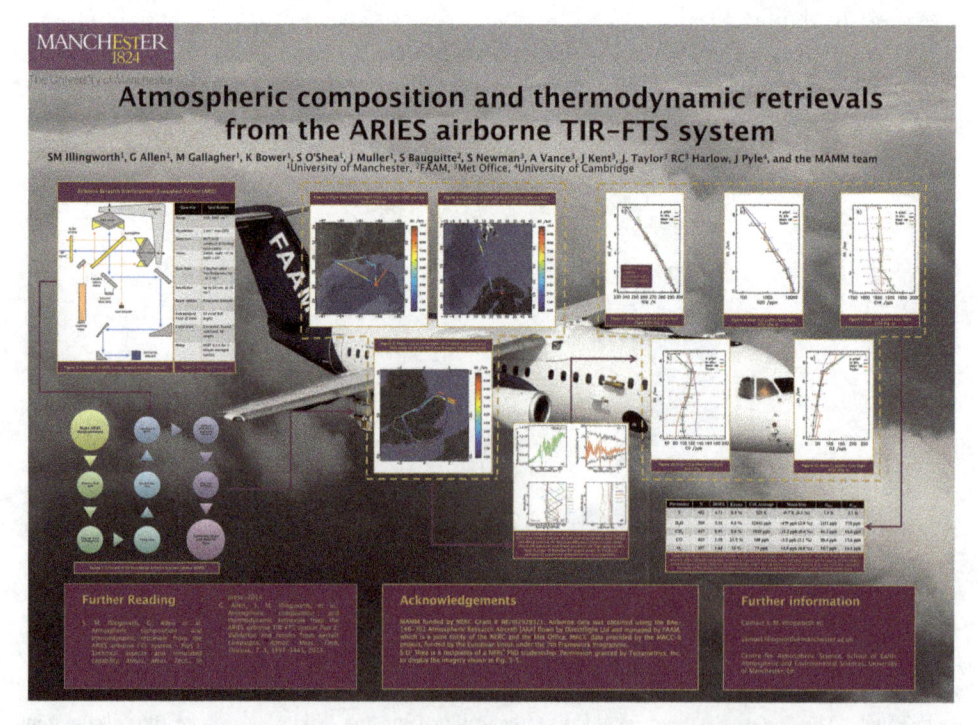

Figure 4.5. A poster displaying the results of a scientific instrument on a research aircraft. Reproduced with the kind permission of The University of Manchester.

Figure 4.6. A poster outlining the development of a tabletop game about heat decarbonisation.

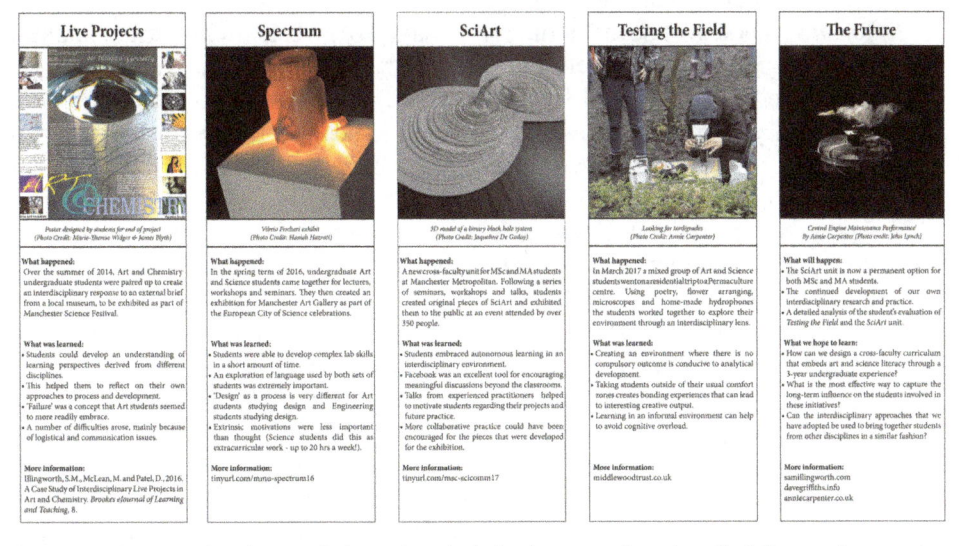

Figure 4.7. A poster based around the pedagogical development of an interdisciplinary science and art programme. Reproduced with the kind permission of Manchester Metropolitan University.

relevant software packages to communicate your central narrative in the most effective, logical, and aesthetically pleasing manner.

4.9 Summary

This chapter has discussed the skills that are needed to be a confident and engaging presenter. In order to be an effective communicator, you need to consider your audience, your narrative, and yourself. Using rhetoric is a powerful way of speaking so that people will listen to what you have to say, and while PowerPoint (or any other presentational software) provides a useful tool, it does not give a presentation for you. When designing a poster or formulating a question for another presenter there are several key points that you need to consider, and if you take the time to participate in all of the exercises in this chapter then you will be well on your way to becoming a first-class orator. The only 'trick' is that practice really does make perfect.

4.10 Further study

The further study in this chapter is related to improving your skills as a presenter, and has been chosen to make you think further about honing a technique that you are comfortable with:

1. **Self-reflection.** Think back to the last presentation that you gave and consider how you dealt with the narrative, the audience, and yourself. Take some time to reflect on whether or not you gave sufficient consideration to each vertex in the Triangle of Effective Communication (figure 4.1). Was there anything that you did particularly well, and what could be improved for your next presentation?
2. **Record yourself.** The next time you have to give a presentation, make a recording of yourself practising the delivery. Then watch it back, and see if you are encompassing all four elements of the SAVE acronym (figure 4.3). Decide if your message has a concise and logical narrative, and if what you are saying is suitable for your intended audience. Then, with these notes in mind practise your presentation a couple more times before filming yourself again. When you watch this next recording see if you have taken on board your own direction; you will probably be surprised at how much you have improved.
3. **Learn from the best.** When you next watch a particularly skilled orator make a public speech (either in person or on the television), try and break down their message into the three different forms of rhetoric. When are they appealing to your emotions (pathos), when are they driving home the logic of the story (logos), and when are they reminding us of their credibility as a speaker (ethos)? You will soon come to realise that many of these skilled orators are experts in using rhetoric for the subtle manipulation of their audience.

4.11 Suggested reading

There are many books and websites dedicated to helping you become a better public speaker, however some of the best are freely available via resources such as the Technology, Entertainment and Design (TED) talks. One of the most helpful of these is a TED talk from the audio expert Julian Treasure, entitled 'How to speak so that people want to listen' [11]. For those of you wanting further guidance on how to design aesthetically pleasing presentations and posters, *Designing Science Presentations: A Visual Guide to Figures, Papers, Slides, Posters, and More* [12] is highly recommended. There are also several short and useful journal articles on designing and delivering successful oral [13, 14] and poster [15] presentations. Finally, if you are interested in finding out more about rhetoric, Aristotle's *The Art of Rhetoric* [6] is the seminal text on the subject; in addition to expanding on the concepts discussed in this chapter, it also offers sound advice on how to survive in love, war, and everything else in-between.

References

[1] Peck E and Dickinson H 2009 *Performing Leadership* (Basingstoke: Macmillan)
[2] Sanderson T The up-goer five text editor https://splasho.com/upgoer5/ (Accessed 16 October 2019)
[3] Munroe R xkcd: Up goer five https://xkcd.com/1133/ (Accessed 16 October 2019)
[4] Nisbet M C and Mooney C 2007 Framing science *Science* **316** 56
[5] Gächter S, Orzen H, Renner E and Starmer C 2009 Are experimental economists prone to framing effects? a natural field experiment *J. Econ. Behav. Org.* **70** 443–6
[6] Aristotle 1991 *The Art of Rhetoric* (London: Penguin)
[7] Prezi https://prezi.com/ (Accessed 16 October 2019)
[8] Kahoot! https://kahoot.com/ (Accessed 16 October 2019)
[9] Poll Everywhere https://polleverywhere.com/ (Accessed 16 October 2019)
[10] QR Code Generator https://qr-code-generator.com/ (Accessed 16 October 2019)
[11] Treasure J How to speak so that people want to listen https://ted.com/talks/julian_treasure_how_to_speak_so_that_people_want_to_listen?language=en (Accessed 16 October 2019)
[12] Carter M 2012 *Designing Science Presentations: A Visual Guide to Figures, Papers, Slides, Posters, and More* (London: Academic)
[13] Collins J 2004 Education techniques for lifelong learning: giving a powerpoint presentation: the art of communicating effectively *Radiographics* **24** 1185–92
[14] Bourne P E 2007 Ten simple rules for making good oral presentations *PLoS Comput. Biol.* **3** e77
[15] Erren T C and Bourne P E 2007 Ten simple rules for a good poster presentation *PLoS Comput. Biol.* **3** e102

Chapter 5

Outreach and public engagement

Science, I maintain, is an absolutely essential tool for any society with a hope of surviving well into the next century with its fundamental values intact—not just science as engaged in by its practitioners, but science understood and embraced by the entire human community. And if the scientists will not bring this about, who will?

—Carl Sagan

5.1 Introduction

So far in this book we have mainly discussed how to develop inward-facing skills for communicating effectively within the scientific community. However, this is only a small section of society. Why then, as scientists, should we consider developing our outward-facing skills to communicate with people from outside of this community? How can we do this effectively? And what's in it for us?

As scientists, we communicate with non-scientists for three main reasons: because we have to, because we want to, and because we should do. The majority of large research grants (see chapter 3) now require a consideration of 'pathways to impact', i.e. the development and delivery of initiatives that will increase the likelihood of potential economic and societal impacts being achieved. In order to develop and deliver these initiatives it is necessary for the scientists involved to fully understand the potential of these impacts. Furthermore, it is largely the taxes paid by the wider society that fund much of our scientific research; they are stakeholders in our successes and failures. *We communicate because we have to.*

For the majority of scientists, the main reason that we are in this profession is because we are passionate about our subject and have a thirst for knowledge, discovery, and truth. There are certainly far more financially rewarding and secure career paths available (see chapter 9), but science is, for the most part, a vocation.

Given our appreciation of the subject, it follows that most of us also enjoy talking about our work and research, not only to other scientists but to friends, family members, and generally anyone with a passing interest in what we do. *We communicate because we want to.*

As discussed in chapter 4, as scientists we have a responsibility to question illogical and misleading arguments and statements of unsubstantiated 'fact'. In doing so we also have a responsibility to help train non-scientists to question the status quo objectively, and to give them the confidence and skills to interrogate failures of truth. It is easy to forget the educational and intellectual privilege that we enjoy as scientists, and that many other people are often not in a similar position. *We communicate because we should do.*

However, scientists are not just obligated to communicate their research and other scientific advances to the rest of society in a one-way flow of information. Rather, they should be aiming to engage non-scientists in meaningful two-way dialogue, helping to create a society in which science is not only better understood, but also challenged, contested, and collaborative. This chapter will provide advice on how such dialogue can be developed in a meaningful way.

Aside from our responsibilities to engage with non-scientists, doing so also helps to develop us as scientists; for example, by improving our communication and organisational skills. Learning to communicate with different and varied audiences is also of direct benefit to how we communicate our research within both the broader scientific community and in university-level teaching. Furthermore, the interpersonal and teamwork skills that we develop when engaging with non-scientists are also essential in a variety of careers (see chapter 9).

Gaining experience in working with non-scientists is an advisable first step before starting to develop your own science communication initiatives, as doing so will help to build your skills and confidence in dealing with a range of audiences across a variety of formats. Many research institutes provide free training and professional development opportunities, and will likely have a large selection of initiatives that you can gain experience with. In addition, some funders and learned societies also offer training to researchers who want to develop science communication initiatives based on their research.

Some scientists may consider themselves to be 'too busy' to engage with non-scientists, or have a misconception that doing so results in little reward for a lot of effort. However, while developing effective science communication initiatives does indeed involve considerable commitment, the benefit for all parties can be considerable. Furthermore, as scientists we have an obligation to help to nurture a society in which scientific understanding extends beyond a basic knowledge of facts, and includes an ability to engage in meaningful discussions about the function and purpose of science. This chapter has been written to provide guidance for achieving these demanding, yet rewarding and necessary aspirations.

5.2 Objectives, audiences, and formats

Science communication is not a new phenomenon. In the UK, the concept of scientists communicating their research findings to non-scientists dates back at least as far as the early 19th century, when scientists such as Michael Faraday and Humphry Davy spent a considerable amount of time and money trying to popularise science. However, as an academic discipline, science communication is a relatively new field, which in the UK has undergone three main stages [1, 2]: scientific literacy, Public Understanding of Science (PUS) and Public Engagement with Science and Technology (PEST). In moving through these three stages, the ideology of the communication of science has developed from a primarily deficit model (in which scientists try to 'fill' gaps in the knowledge of the public) into one which encourages two-way dialogue between scientists and non-scientists.

Science communication as an academic discipline covers a broad range of topics [3, 4]. However, as the majority of research into the relationships between scientific knowledge, technological systems, and society tends to be done by STS (Science and Technology Studies) scholars, perhaps too little of the current recommendations and accepted best practices are communicated to scientists who are actively engaging with (or planning to engage with) non-scientists [5, 6]. Furthermore, the field of science communication tends to defy any singular definition [7], with any such attempt needing to reflect: the variety of formats in which such initiatives can be implemented [8], the wide spectrum of target audiences [9], and the range of objectives [10]. This multi-faceted nature is also illustrated by the many terms that are used when discussing outward-facing science communication, such as: widening participation, knowledge exchange, public engagement, and outreach [11]. As well as institutional and national biases towards the 'correct use' of these terminologies there often exist personal nuances in terms of their interpretation, depending on how individuals perceive science communication to fit into their individual research practices, and beyond.

Based on the current science communication literature, and our own experiences, the following broad definitions are offered:

Outreach: a one-way discourse, in which scientists communicate their research to non-scientists.

Public Engagement: a two-way dialogue, in which scientists converse with non-scientists in a mutually beneficial manner.

Widening Participation: any initiative that engages with social groups under-represented in higher education, in order to encourage them to attend university.

Knowledge Exchange: any initiative that involves engagement with businesses, public organisations, and third sector organisations (e.g. charities).

We acknowledge that there is still some overlap between these definitions. For example, a science talk at a local school given by a UK-based university researcher might be classed as being outreach, widening participation, and knowledge exchange. In this example, the researcher might classify the initiative as outreach, the university's widening participation team may catalogue it as widening participation, and the university's knowledge exchange offices could acknowledge it in their records for UK Research and Innovation.

Widening participation and knowledge exchange as defined above are beyond the scope of this chapter, which will instead focus on outreach and public engagement, these being the most likely to be encountered by the majority of scientists. For brevity, we will use the term 'science communication initiatives' to refer to both outreach and public engagement initiatives unless otherwise stated.

Exercise: planning your initiative

When you are developing any science communication initiative you should begin by asking yourself these three questions:

What is your objective? For example, do you want to raise awareness of the importance of diversity in scientific research? Are you interested in finding out the opinions of a local community group to inform your work on flood risk mitigation strategies?

Who is your audience? How does this help you to achieve your objective, and how will you reach them? For example, if your objective is to raise awareness of air pollution amongst pensioners then how will you engage with this community?

What format will you use? This needs to enable you to both achieve your objective and be appropriate for your target audience. For example, if you want to engage with local farmers in order to better understand the soil quality of arable farmland in the region, then a series of workshops might be more conducive than a one-off science talk.

Your objective is what will drive your science communication initiative, and as such it needs to be clear and achievable. You might also have a particularly aspirational long-term objective that can then be broken down into several short-term objectives. For example, your long-term objective might be for the health effects of air pollution to be more fully incorporated into your country's school curriculum. However, in order to achieve this, your short-term objectives might be to develop a game that is used in 30 local schools to raise awareness of the subject, and the organisation of five panel debates with educationalists and policymakers to discuss the potential for re-designing the curriculum. These objectives will depend entirely on what you want to achieve, and so the remainder of this chapter will instead focus on providing support for the other two questions, i.e. how do you engage with suitable audiences, and what are the practicalities of the different formats that can be adopted.

5.3 Different publics

When thinking about which audience you want to engage with, the term 'general public' or 'lay audience' is somewhat misleading, as in reality there are many publics [9]. Simply targeting an audience that are not scientists is ineffectual, as 'not being a scientist' is not generally an identity, behaviour, or characteristic that people tend to identify and group themselves according to [12]. In developing your science communication initiative, it is essential that you consider which publics you intend to target, and why. In some instances, you might not have a choice, as your audience may be pre-determined by a larger initiative of which yours is only a part. In which case, have you considered if this larger initiative is a suitable platform for achieving your objectives? For example, if your objective is to raise awareness of climate change amongst local business leaders, then organising a panel debate at a local primary school (during the working week) might not represent the most effectual targeting of a suitable audience.

When determining which audience to engage with, try to think beyond previously-engaged audiences. The reasons for this are twofold: firstly, working with familiar audiences risks trapping you inside an echo chamber, and secondly there are many audiences that are underserved and under-represented by both science and science communication initiatives, and as ethical scientists (see chapter 9) we have a responsibility to engage these audiences.

An echo chamber is an environment in which a person encounters only beliefs or opinions that resonate with their own, so that existing views are reinforced and alternative ideas are not considered. As scientists we should be breaking out of these echo chambers, moving away from the same, traditionally engaged audiences. Instead, we should be enabling more effective communication with publics that have a diverse range of demographic, socio-structural, and value-based characteristics [13]. Working with these underserved and under-heard communities presents barriers and challenges, but it also creates many opportunities for both advancing scientific research and re-considering what is meant by meaningful impact [14].

But how do you engage with these audiences? Begin by developing a relationship with a member of the community that you want to work with, ideally someone who is in a position of responsibility; for example, if you are interested in working with a local youth group then you might first make contact with one of the adult leaders. Developing this relationship may involve emails, phone calls, and meetings over a relatively long period of time before you are able to deliver any initiative, as it will be necessary to build trust, especially if the community that you intend to work with includes people who might be considered to be vulnerable. When developing your initiative, involve your contact in the community, as they will have a better understanding of how you can tailor your plans to the relative needs and experiences of the target audience. Also, think carefully about where the initiative will take place; while research institutes are often convenient, 'safe ground' for the scientists involved, for many publics they are inaccessible locations that present several physical and psychological barriers. Finally, in thinking about which communities to develop your science communication initiative for, have you considered the communities to which you already belong. For example, do you volunteer at a local charity? Are you a member of a sports club? Do you host a weekly games evening? These are all communities which you might consider developing an initiative for, helping to refine your objectives in the process. The benefit of engaging with such communities is that you already have a relationship with existing members, and will be naturally sensitive to their needs and experiences.

In developing any science communication initiative that involves scientists and non-scientists, you need to give thought to how you can level any so-called 'hierarchies of intellect'. These arise when there is a perception that one of the parties is an expert and the other is not [15], and can hinder meaningful discussions. Scientists might be experts in a particular topic, but non-scientists are experts in their own personal and professional capacities, and this expertise should be encouraged and highlighted in order to facilitate a more conducive environment. For example, if you want to develop a science communication initiative to find out what a rural community think about genetically modified foods, then create a platform in which non-scientists are asked to share their own expertise and knowledge, and where they are treated in a similar manner to the scientific 'experts' who are asked to do the same.

Once you have determined your audience and how you will reach them, you need to think about how you will frame your initiative. Framing was introduced in chapter 4 when discussing the importance of understanding your audience for science presentations. The advice that was provided there still holds, i.e. that understanding the needs and experiences of your audience is key to effectively framing any science communication initiative. Furthermore, in framing the way in which to discuss certain scientific topics, we also need to ensure that we avoid promoting false expectations [16], and to behave as ethical scientists throughout the process.

5.4 Working with children

The advice that has been given so far in this chapter has been of a purposefully general nature, as it is not possible to provide specific advice for all of the various publics that you may encounter. However, given that for many scientists one of the first outward-facing publics that they will work with may be children, we thought it beneficial to provide more detailed advice for developing and delivering science communication initiatives for this particular audience. In most instances, these initiatives will take the form of outreach (i.e. one-way discourses, in which scientists communicate their research to non-scientists), although this need not always be the case.

Working with children can be an extremely rewarding and enjoyable experience. However, it can also be demanding, difficult, and at times disheartening. It is arrogant to assume that you will simply be able to walk into a classroom, or a more informal setting, and instantly command the room because you are *a scientist*. Before developing and delivering an initiative aimed at children it is advisable to first get some relevant training and experience.

The STEM (Science, Technology, Engineering, and Maths) Ambassadors [17] are a national collective of volunteers who are dedicated to providing science communication opportunities in STEM subjects across the UK. In addition to providing training and advice, they also have a list of ready-made initiatives that are being run by schools or other organisations, which you can participate in to receive science communication experience.

If you are working with children in the UK then you need to have a Disclosure and Barring Service (DBS) check, to make sure that you are fit to work with children. Other countries have similar requirements, and there is normally a small fee attached to having one of these checks performed, but your research institute will usually pay this for you if you are delivering initiatives on their behalf. However, a word of caution, even if you have an up-to-date DBS certificate: **you should never be left alone in a room with a child or group of children.**

Having a teacher or a guardian in the room with you at all times is a necessity, and guards against any potential claims of malpractice. Having a teacher present will also make it far easier for you to interact with children as they can help to introduce you and to control the environment, and can also work with you to deliver the initiative through careful co-design (see section 5.4.1).

The age of the children that you are working with will, to a large extent, determine the type of initiative that you develop and/or deliver. It is wrong to generalise and think that all teenagers are moody, and will have no interest or passion for science. However, some of them may have a negative attitude towards science because of a lack of engagement, poor teaching, or even previously ineffective science communication initiatives. In contrast, working with younger children can be a liberating and exhilarating experience. They are yet to develop the cynicism and awkwardness that can sometimes make engaging with older children so frustrating. Though be warned, the ebullience of this audience can also present problems in terms of behaviour and exhaustion. When working with any children, the same advice that was discussed in chapter 4 applies: do not attempt to patronise your audience; instead, ask them about their needs and experience and develop/deliver your initiative accordingly. Carefully developed science communication initiatives can engage and empower children, instilling in them a love of science at this early and impressionable age.

Exercise: what does a child know?

It is easy to forget that as a scientist you know (and are surrounded by colleagues who know) a lot of information about your research, its related discipline(s), and science more generally. What you consider to be common knowledge might in reality be highly specialised information, especially to a child.

The next time that you have the opportunity to speak to a young child in an informal and supervised location, ask them what they know about science. Start off with questions that are quite general (What does a scientist do? What is physics?), and then begin to specialise (What is acceleration? What is gravity?). You will probably be surprised to find out what many children do (and do not) know, and you should use this to help structure your future science communication initiatives for this audience.

5.4.1 Children in a formal environment

The majority of science communication initiatives that involve children in a formal environment (i.e. the school classroom) can be classified as outreach (see section 5.2). The purpose of these initiatives is usually to engage with a group of schoolchildren about a particular area of science, to raise aspirations, and to re-normalise who scientists are and what they do (see chapter 9). However, rather than focusing on the children's supposed lack of knowledge, an approach based around the understanding of the learner(s) and the learning process should instead be considered.

Focussing engagement around the experiences and knowledge of schoolchildren involves a detailed understanding of the school curriculum, and an in-depth awareness of the needs and abilities of every child in the classroom. This is a lengthy process that cannot be fast-tracked, nor is there a need to do so when the children's teachers can provide this information. As such, it is advisable to involve a teacher in the development process as early as possible. Their knowledge of the curriculum and of general learning behaviours within the school environment will ensure that your initiatives are effective in engaging *their* students. They will also be able to provide constructive feedback with regards to what will (and will not) work in their teaching environments. Teachers should also be able to assist with basic logistics such as room setup, and will be able to ensure that the class is grouped (where necessary) to avoid disruption.

In instances when you have experience of successfully running a particular science communication initiative in a formal classroom environment, it is still good practice to engage with the teachers of any class prior to your delivery. It is wrong to assume that what works for one group of schoolchildren will work for another, and by providing a basic summary to the teacher beforehand, they will be able to give feedback as to what will and will not be engaging for their students.

Here are five further pieces of advice to consider when working with school-children in a formal environment:
1. **The children are not your friends**. They are there to learn, and while they can have fun during the process, boundaries need to be established.
2. **Stick to time.** Schoolchildren will not thank you for eating into their breaks. For initiatives that take place in the afternoon, make sure that you finish with plenty of time to spare, as many of the children will have buses to catch or parents waiting to take them home.
3. **You know more science than they do**. A common fear of many scientists working in schools is that they will be 'caught out' on an area of science that they do not know. Ninety-nine times out of one hundred you will be able to answer any of the questions that you are asked. And for that one hundredth time, simply commend the questioner, and tell them that you will have to conduct some research before reporting back to them; alternatively, you could offer to work with the children to find out the question together. Admitting your lack of knowledge might also help to empower the teacher (who might not have a science background) when fielding difficult science questions from the children in the future.
4. **Expect the unexpected.** Be prepared to answer questions about your life as a scientist, and indeed your life in general. Young children in particular will be fascinated about what it is like to be a scientist, which exotic locations you have visited in your fieldwork, and how often you get to use robots.
5. **Don't get disheartened:** On occasions things will not go as planned. This may be for a number of reasons: the children, the facilities, the alignment of the planets, etc. Do not dwell on any negative experiences, instead reflect on what went wrong and how it can be used to improve future initiatives (see section 5.9 for a further discussion of reflection and evaluation).

Exercise: develop an outreach initiative for the classroom

Follow these steps and devise an initiative to discuss your research with schoolchildren in a formal learning environment:
1. What is your objective? Do you want to raise awareness of a particular subject? Introduce the class to a famous scientist? Better understand what they know about particle physics?
2. What format is most suitable for achieving your objective with this audience? Is it via a short presentation, a series of demonstrations, some hands-on experiments, or something more creative?
3. How does this tie in to the school curriculum? Your initiative will be more effective if the topics that you are covering can be linked to the curriculum. This is especially true for more mature schoolchildren, where classroom time is often on a tight schedule.
4. Run your ideas past a teacher. They will be able to advise what will and will not work in their classroom, and will also be able to help with linking your plans to the taught curriculum.

5. Beta-test your initiative. Aim to have at least a couple of dry runs before taking the initiative into a school, as this will help you to iron out any issues beforehand. Undergraduate and postgraduate students are great for helping at this stage.
6. Trial your initiative. Get in contact with the teacher that you spoke to in the development process, and see if they are willing to let you try out your initiative in their class.
7. Reflect on the trial. What went wrong and what went right? Ask for feedback from the teacher and their class (see section 5.9), and also from the people that were involved in the beta-testing process. How can you use this feedback to improve your initiative, and do you need any further support and/or resources to better implement it?

5.4.2 Children in an informal environment

Learning does not just take place in the classroom. There are many different environments outside of school where children can continue to learn about science in a more informal setting, including: museums, science centres, and even zoos. However, informal science education is not just defined by learning that takes place outside of the classroom, but rather as something that is self-motivated and guided by the learner's needs and interests [18].

Large science initiatives for children often take place in these informal settings, and include science festivals, science fairs, and public lectures. For example, the Royal Institution Christmas Lectures in the UK have been running since 1825 and are aimed at a mainly teenage audience, taking place at the Royal Institution in London each year [19]. Informal science communication initiatives such as these have been shown to foster a strong commitment to science and science learning, and to have a strong impact on future career choices [20].

When running an initiative in an informal environment that is aimed primarily at schoolchildren, take account of the following:

1. A teacher might not accompany the children; instead a guardian might be present, or they may be unattended. In any case, the children will behave differently outside of the school environment. They may feel less awkward, but similarly there may be behavioural issues that need to be kept in check without the presence of a teacher. In these informal environments it is just as imperative that you are never left alone with any children.
2. If your initiative is not part of a larger science festival, or is alongside other initiatives that are not science-themed, then your participants may not be expecting to do any science. These 'science by stealth' opportunities [21] are an effective way of reaching new audiences, that might not otherwise seek out science-specific initiatives.
3. There might be a larger or a smaller influx of people than you were expecting. Plan for both eventualities, especially when arranging the number of scientists that will be involved. Where possible, have several activities that are flexible in the number of people they involve and the time they take to run; doing so will mean that you can engage both small and large audiences accordingly.

5.5 Different formats

Just as there are a diverse range of publics, there are also a large variety of formats that can be used to achieve your objectives and engage your target audience. In this section several different formats, and the practicalities for running these, are discussed. While this is by no means an exhaustive list, the formats presented here have been chosen to demonstrate the wide variety that is available.

Some of the formats discussed below might be considered to be examples of outreach (one-way communication), some are public engagement (two-way communication), and some have the flexibility to be both. Over the past couple of decades, research in the field of science communication has generally tended to recommend public engagement formats as being more effective than outreach in engaging different publics [4]. This is largely because outreach is often associated with a deficit model of engagement, which has in turn been heavily criticised as being ineffectual, over-simplified, and derisory, in assuming that non-scientists are 'deficient' and scientists are 'sufficient' in knowledge [22]. However, while meaningful dialogue over science-related issues is essential for the development of science, and society more generally, providing reliable information in an accessible way is often an essential prerequisite for this to occur [23]. Furthermore, a gain in knowledge can have positive impacts on people's attitudes depending on their contexts and pre-knowledge [22]. If done correctly, outreach initiatives that are one-directional in format can still be extremely effective in achieving objectives and engaging audiences. We need only look at the impact that nature documentaries, such as the *Blue Planet* series hosted by David Attenborough, have had to see evidence of this [24].

The following introduction to these selected formats should help you to think about how to develop your own science communication initiative, with both your objectives and your audience in mind. Section 5.11 provides examples of some successful science communication initiatives that have made use of some of these formats.

Science talks
The most standard form of outreach is a science talk. This may consist of a lecture-style talk with an accompanying Q&A session, or a more informal discussion such as those hosted by a Café Scientifique [25]. Whatever the setting, the advice provided in chapter 4 still applies: consider your narrative, your audience, and yourself. Also, just because you are not speaking at an international scientific conference, do not assume that there are no experts in the room. Instead, try to find out who your audience will be, so that you can avoid either overestimating their knowledge or underestimating their intelligence. The advice that was given in chapter 4 with regards to preparation is also appropriate here: find out what AV equipment is available, and try to arrange a practice session or sound check in advance if possible.

Panel discussion
A panel discussion is an effective way of showcasing a variety of different opinions and knowledge surrounding a certain topic. They also help to demonstrate to the audience that science is a varied and much-debated topic, in which there are sometimes quite

fierce and contrasting views. If taking part in a panel discussion, find out in advance about the format (round table, open Q&A, short presentations, etc) and also your fellow panellists, and their attitudes regarding the topic(s). If you are organising a panel discussion then choose a topic that is relevant to the intended audience, and invite a diverse selection of panellists (not just scientists), who can represent different points of view. If you are recording the panel discussion then get explicit permission from the panellists, and ensure that they are aware of how it will be shared (e.g. streamed via social media, or hosted on an institution's webpage). Picking a chairperson who can both keep to time and ensure that all voices are heard is also essential.

Science busking
Science busking involves capturing people's attention in a public space using the 'magic of science'. For example, you might make a cloud using only a bottle of water and a lit match, or demonstrate surface properties by putting a wooden kebab stick through a balloon without it bursting. When done properly, this can be an effective, enchanting, and innovative way of engaging a potentially large group of people. Like other forms of street performance, there is a definite skill in engaging an audience, and inspiring people to want to approach you to ask questions. If you are interested in finding out more about science busking, then Science Made Simple and the British Science Association have created a useful resource, which includes a selection of science busking activities that are suitable for all audiences [26].

Book clubs
Setting up and hosting a book club provides an innovative way of discussing science in an accessible and engaging format. If you plan on running a book club then it helps to have an overarching theme that is not too broad; for example, books that involve 'time travel,' rather than 'science' in general. Meeting once a month will give people enough time to read the selected title, and choosing books that are readily available from local libraries will help to keep the costs down. It is also recommended that you plan out a number of books in advance, and that each member of the group gets the opportunity to select a book as well as to take part in the discussions. You might also consider setting up a digital book club, in which members meet on social media (Twitter is ideal for this) at an allotted time. If you take this approach, then account for different time zones if you want to include a more diverse audience.

Workshops
We are using the term workshop here as a catch-all term that involves working with an audience in order to discuss and deliberate a specific topic of interest. This might be a one-hour meeting over tea and biscuits in a local community centre in which participants are invited to chat to scientists about their knowledge of the solar system, or a series of initiatives in which scientists and non-scientists are asked to brainstorm ideas for future clean energy solutions to present to local policymakers. Whatever the format of your workshop, they should have a clear objective, and be framed for a specific audience; they should also be conducted in a way which enables the participants to feel safe, and where all voices can be heard and respected.

Citizens' juries
A citizens' jury is a special type of workshop; a specific method of deliberation in which a small group of people (typically 10 to 20) come together to discuss a well-framed question or issue, over a time period of two to seven days. The jury members are selected to be representative of the target audience, and the aim of the jury is to allow non-scientists to meaningfully discuss, in detail, a topic that tends to be either controversial or of deep societal significance. Developing and delivering a citizen's jury is not something that should be taken lightly, as they require large amounts of resources, in terms of both time and money. However, they can create a platform which genuinely involves the participants, granting them ownership and agency of the process. Involve, the UK's leading public participation charity, have a large variety of resources that can help you to plan a citizens' jury, including detailed explanations of suitable methods and successful case studies to draw from [16].

Whatever format you decide upon, it is vital that you consider the ethical implications of your initiative. If you are planning on conducting any research, or carrying out an evaluation which involves collecting personal data from the participants (see section 5.9), then you should seek ethical clearance from your research institute. Even if you are not collecting any data from the participants you should still think carefully about the repercussions of your proposed format. For example, if you are talking to a group of elderly people about the latest medical research on dementia, then be sensitive to the effect that this may have on some of the audience members. Similarly, if you are planning on discussing anything that others might perceive to be upsetting or offensive, then signpost this with appropriate trigger warnings. Developing your science communication initiative with members of your intended audience will help you to identify when and where such incidents may occur.

5.6 Citizen science

Citizen science is a popular example of a public engagement format. In essence it is a type of collaborative research that involves members of society (or citizens) in actively collecting, generating, and in some instances analysing data.

There are many examples of citizen science projects, but one of the most well-known is Galaxy Zoo [27], an online series of projects which invites participants to classify different types of galaxies according to their structure; the human eye being better equipped at making these distinctions than a machine. There have been many versions of Galaxy Zoo, with Galaxy Zoo 1 (which ran from 2007 to 2009) receiving more than 50 million classifications from over 150 000 people in just year one of the project.

Another popular citizen science project is Old Weather [28], which aims to help scientists recover weather observations made by US ships since the mid-19th century, by enlisting citizens to digitalise old transcriptions recorded in ship logbooks. Such information ultimately improves the collective knowledge of past environmental conditions, with a better understanding of these past occurrences leading to an improvement in modelling future weather patterns.

There are also a number of citizen science programmes that actively source data directly from members of the public. For example, the Community Collaborative Rain, Hail & Snow Network [29] is a non-profit, community-based network of volunteers who measure and map precipitation using low-cost measurement tools with an interactive website. The project started in Colorado in 1998 and now has networks across the United States and Canada, involving thousands of volunteers, and making it the largest provider of daily precipitation observation in North America.

The main objection to these types of citizen science projects are that they are potentially tantamount to free labour, with scientists relying on non-scientists to collect and/or analyse vast swaths of data. While there are many incentives for performing these tasks (such as prizes, badges, and general kudos), it tends not to be the citizens whose names appear on the associated research publications and/or grant applications.

If you are thinking of developing a citizen science initiative then make sure that the citizens you recruit are properly recognised, and where possible involved throughout the whole process. As an example, the UK Community Rain Network, in which children from across the UK monitored precipitation using home-made rain gauges, acknowledged all of the participating citizens in the subsequent journal publication [30].

Overall, citizen science projects are becoming an increasingly popular means by which to engage the public, while also benefiting scientific research, especially given the growing ubiquity of social media and other communications platforms (see chapter 7). However, there is a need to actively involve the participants in these projects, and to ensure that they receive the appropriate acknowledgements; otherwise scientists run the risk of treating their new colleagues as nothing more than second-class citizens.

If you are interested in setting up your own citizen science project, then the Natural Environmental Research Council and the Natural History Museum have produced a very useful guide on how to do this both effectively and ethically [31].

5.7 Funding

After determining how your science communication initiative will meet the needs of your objective and your audience, you need to consider how to finance it. Even the most basic of initiatives will have some consumable costs, while larger initiatives will also have to account for travel and venue hire, plus maybe even a contribution to the salary of those involved. Here are five potential revenue streams for you to consider:

1. **Science communication funding.** The National Co-ordinating Centre for Public Engagement (NCCPE) have a useful resource on their website which lists most of the funding grants available for science communication in STEM in the UK [32]. When applying for one of these grants, follow the advice given in chapter 3; in addition, getting match funding from one of the other sources on this list will greatly strengthen your application.
2. **Universities.** Almost all universities have a widening participation team, with many also having people dedicated to co-ordinating science communication initiatives. Find out who these people are, and ask them for advice when developing your initiative. Furthermore, they might have access to funds that you can utilise, especially if your plans align with the University's wider strategy.
3. **Existing research grants.** As discussed in section 5.1, most large research grants must now demonstrate 'pathways to impact', i.e. they must show how they are making a conscious effort to inform society of the research that they are doing, and the relevance that this has to the wider community. Funds will normally have been set aside to do this, and therefore represent a potential revenue stream for future initiatives.
4. **Local councils.** Your local council will have certain allocations of funds that they must use to help improve engagement with the local population. They are also a very useful source in terms of school and community contacts, and will often have venues that they can offer for use at a reduced rate.
5. **Learned societies.** Most academic or scientific societies offer some kind of support for science communication initiatives, both through formal grants, and also via development funds for associated members and fellows. For

example, the Institute of Physics' Public Engagement Grant Scheme provides funding to individuals and organisations running physics-based science communication initiatives in the UK and Ireland.

5.8 Advertising

With your science communication initiative fully developed, tested, and funded, how do you ensure that there is an audience? If you are doing an outreach initiative that involves going into schools, or having schoolchildren come to you, then it is necessary for you to make prior arrangements with the schools that are involved. As discussed in section 5.3, if you are working with traditionally underserved audiences, then any science communication initiative should be developed alongside a member of this community, who should be able to help advise on how best to locate, schedule, and advertise your initiative in order to cater for this audience. We would also advise taking this approach for all publics, as working with representatives from your target audience will help to identify the most effective way of reaching them.

Mailing lists are a useful way of engaging with a specific (but often largely academic) audience. For example, the PSCI-COMM (psci-com@jiscmail.ac.uk) and PCST (network@lists.pcst.co) email discussion lists reach a large number of people with an interest in science communication, and people on these lists often recommend other local interest groups that might also be interested in your initiatives. Personalised emails to contacts who you know might be interested (or know someone who might be interested) in your forthcoming initiative can also be an effective advertising strategy. On the other hand, generic emails that are sent out to a large group of people will probably not be opened by the majority of the recipients.

It is often worth contacting local newspapers and magazines, as well as international publications with regional offices and initiatives listings, such as *Time Out* magazine. In addition to offering paid advertisements, many of these publications also run free listings, in both online and print formats. Posters and flyers can also help to raise awareness for your initiative, especially if put up in locations that your intended audience frequents, and in places where they cannot help but look at them, such as in elevators or toilet cubicles.

Social media (see chapter 7) can raise awareness amongst a potentially very large group of your target audience in a relatively short amount of time. If your initiative is aimed at a particular audience (e.g. amateur astronomers), or if it is related to a local or global initiative (e.g. World AIDS Day), then you should also be sharing information about your initiative with the people who control the various social media channels of any associated organisations. Learned societies and other organisations that are somehow related to the topic of your initiative should also be informed in advance, as should any funders.

The logistical side of ticketing has been made much easier with the advent of free online tools such as Eventbrite. Experience seems to suggest that as a rule of thumb, for free and ticketed initiatives an attrition rate of about 30%–50% is the norm. This can be especially annoying for people who are unable to attend something that is 'sold out', only to find out later that only half the people with tickets turned up. To account for this, it is good practice to have a reserve list, or to consider introducing a small charge to encourage attendance; it is amazing how a couple of pounds/euros/dollars can incentivise people to attend. There is also an argument to be made that charging even a relatively small amount for a ticket gives inferred value that might actually increase attendance rates. Despite the software available to help with the logistical side of things, ticketing is a fine art that benefits from experience, and sadly one for which there is no one-size-fits-all solution.

"WE'RE GONNA NEED A BIGGER BOAT."

5.9 Evaluation

Failing to properly evaluate any science communication initiative means that you will be unlikely to fully assess if you have successfully met your objectives. As an absolute minimum you should be recording the number of participants who attend your initiatives, both for your own records and those of any associated research institute or external funder. A short personal summary of each initiative is also good practice, as by recording your experiences and reflecting on these (see chapter 9 for a discussion of formal models of reflection) you will be able to make improvements.

In order to fully assess the extent to which you met the outcomes of your science communication initiative, you should try to get detailed feedback from both your audience and also any colleagues who helped deliver the initiative. In constructing a feedback survey think carefully about the data that you want to receive from the participants. Many surveys ask for detailed demographic information, but if you are not going to use this information, then there is no need to ask for it. For example, it might be conducive to ask the participants 'what they enjoyed' or to 'rate the accessibility of an initiative from 1 to 5 (1 being very poor and 5 being very good)'.

However, unless you are conducting research into how gender or age might influence people's attitudes, or you are interested in better understating the demographics of your audience, then there is no need to ask people to provide this information. Finally, please be considerate in the layout and phrasing of your surveys. For example, if you need to ask people what gender they associate with then leave an open space for them to fill in, rather than simply presenting a binary option or a multiple choice of 'male/female/other'.

Google Forms [33] is a useful resource for managing feedback, as it can be used to both create and distribute a variety of surveys, using a dedicated link that can be emailed to participants. If you are worried about people not following up after the initiative then you can either ask them to fill in the survey on their electronic devices before they leave, or print out some copies for them to fill out, which can then be uploaded to Google Forms at a later date. Google Forms also has some basic analytical tools for evaluating the responses, and you can export these into a spreadsheet for further analysis.

One of the major drawbacks of asking participants to fill in a long and cumbersome survey is that it can sometimes detract from what they have just experienced. Where possible, see if you can incorporate any opportunities for feedback into your initiative, and if you can make the process innovative and enjoyable. For example, the feedback form shown in figure 5.1 was used to evaluate a science talk about the geographies of light and dark. These forms were printed out on two sides of A5 card, and were handed to participants at the end of the talk, along with pencils and pens. This resulted in a fun and easy method of giving feedback, which was also quick and enjoyable to analyse.

The feedback that has been discussed thus far is useful for determining the extent to which people enjoyed your science communication initiative, and is helpful for assessing how to improve future initiatives. However, in order to fully assess the success of these initiatives, i.e. the extent to which they have achieved the desired objectives, a more considered approach is required. Figure 5.2 is a pictorial representation of the scientific process: you begin with a hypothesis, design an experiment to test it, carry out those tests, and based on the outcomes of these tests you either accept the original hypothesis or adjust it and repeat the cycle. This is a process which, as scientists, we carry out on a daily, sometimes hourly basis, and yet when it comes to evaluating our science communication initiatives, the majority of us are guilty of forgetting our scientific training, resulting in a lack of meaningful evaluation.

As an example, suppose that you are developing an outreach initiative to raise awareness of subject X for a group of schoolchildren between the ages of 10 and 12. In this instance, the hypothesis would be that 'this initiative raises the awareness of subject X amongst schoolchildren between the ages of 10 and 12.' However, it is not possible to test this hypothesis without first assessing the level of awareness that these students had about subject X prior to your intervention. In this instance, the evaluation process needs to begin before you even set foot in the classroom.

Assessing any prior level of understanding or awareness needn't be overcomplicated; if for example your initiative is aimed at raising an audience's awareness of

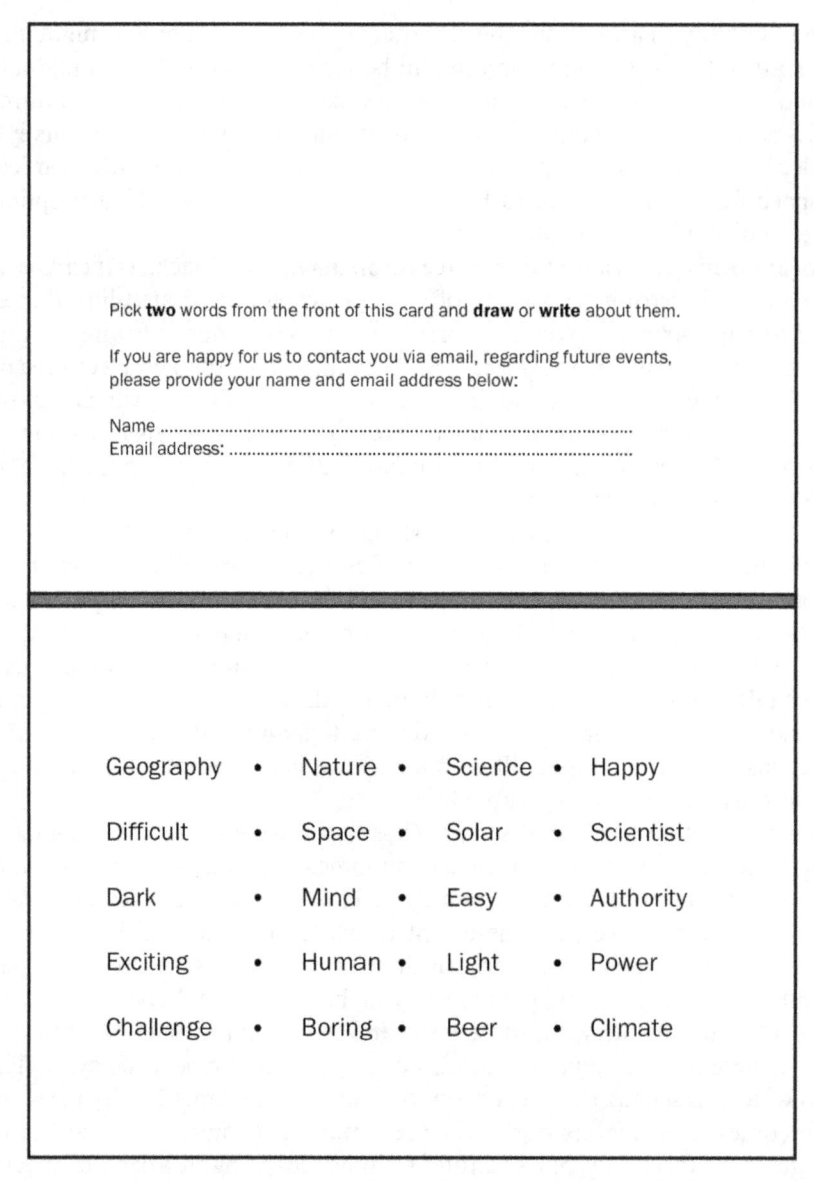

```
Pick two words from the front of this card and draw or write about them.

If you are happy for us to contact you via email, regarding future events,
please provide your name and email address below:

Name ........................................................................................................
Email address: .......................................................................................
```

Geography	Nature	Science	Happy
Difficult	Space	Solar	Scientist
Dark	Mind	Easy	Authority
Exciting	Human	Light	Power
Challenge	Boring	Beer	Climate

Figure 5.1. A sample feedback form for a science talk. Using feedback forms that are fun to fill out can help to enhance the audience's overall experience of your initiative.

global warming, then their initial familiarity with the subject could be assessed by asking them questions such as: what is global warming? What causes global warming? What can be done to reduce global warming? These same questions could then be asked after the initiative, and your hypothesis could then be better tested based on a comparison of the participants' pre- and post-understanding of the subject. Monitoring any lasting effects, for example by asking the audience to complete a short questionnaire six months later, is an even more effective way to

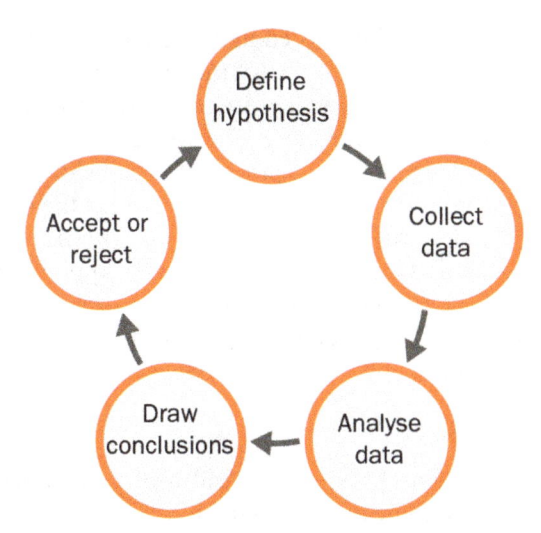

Figure 5.2. A circular diagram representing the scientific process. As scientists we follow this approach when we conduct our scientific research; we should adopt a similar approach when we evaluate our science communication initiatives.

assess the impact that your initiative has had; such an approach is referred to as longitudinal evaluation.

This particular approach to assessing the prior and posterior level of understanding can, for some, be overly reminiscent of 'assessment', resulting in negative attitudes about your initiative. In such cases it might be better to adopt a more informal 'focus group' approach, where participants are encouraged to chat about subject X both before and after the initiative, with their comments and remarks recorded and later analysed by yourself and your colleagues.

Any detailed evaluation that involves capturing information and attitudes from participants, should be designed using the advice that was provided for creating a feedback survey. Furthermore, if the evaluation of these responses is to be used for any report or future publication (see below) then informed consent must be sought from all of the participants. It is also recommended that you seek ethical approval from your research institute, and that you design appropriate participation information and consent forms that outline how and where this information will be used (and stored), and which provide details of how the participants can get in touch if they later wish to have their responses redacted. It should also be made explicitly clear that taking part in this evaluation is not a prerequisite for participation in the initiative itself. When working with schoolchildren and other potentially vulnerable audiences, special care must be taken to ensure that they fully understand the implications of their participation, and where appropriate explicit and informed consent from a guardian should be sought.

Adopting this rigorous approach to evaluating your science communication initiatives means that you might consider publishing the process and its outcomes, alongside what these findings mean for the wider scientific community, in a suitable journal such as one of those listed in chapter 1. Publishing original findings in peer-reviewed journals can help to justify the legitimacy of any science communication initiative to your line manager, supervisor, or external funding body. Furthermore, such publications can help to develop your reputation, while also helping to advance knowledge and best practice in the field.

When preparing a manuscript for one of these journals the same advice that was provided in chapter 2 holds true. Journals such as *Geoscience Communication* also provide support for scientists who are thinking about applying the same scientific rigour to their science communication initiatives as they do to their scientific research, but who might be inexperienced in preparing such manuscripts in the field of science communication.

5.10 Initiative checklist

The table below represents a checklist that will help you to deliver, develop, and evaluate your science communication initiatives.

	Have you thought about...?	Check
	YOUR OVERALL STRATEGY	
Objective	**What do you want to achieve?** Make sure that you have achievable, and measurable objectives, focussing on both the short- and the long-term.	
Audience	**Who do you want to target?** There are many different publics, so think carefully about your objective, and why you want to work with this audience in particular.	
Format	**How will this achieve your objective?** Pick a format that is suitable to both your objectives and your audience, and where possible discuss this in advance with a member of your intended audience.	
	DEVELOPMENT	
Development	**Are you developing an initiative for schoolchildren?** If so, then work with a schoolteacher in the development process. Doing so will ensure that your initiative is suitable for the students and their school curriculum.	
	Are you developing an initiative for a specific public? If so, then work with a member of this public in the development process, as this will help to ensure that your initiative is suitable and accessible for your proposed audience	

(Continued)

(Continued)

	Have you thought about...?	Check
Funding	**How will you fund your initiative?** Look for internal and external funding schemes that you can apply for, and remember to include transport and refreshment costs.	
Advertising	**How will you advertise your initiative?** Use targeted email and social media advertising, and work with your intended audience to build a supportive and trusting relationship.	
Staff	**Have you got enough facilitators?** Involve these people throughout the development process, checking that they have permission from their line managers. Also, consider if these facilitators are 'volunteers', or if they require payment for their time and expertise.	
	Have you provided suitable training? Anyone who is working with under-18s needs to be briefed on safe and appropriate ways of working with young people. The same goes for when working with any potentially vulnerable audiences.	
	Is there appropriate identification for the facilitators? Wearing badges/t-shirts/fleeces etc can make it easier for participants to get help.	
Insurance	**Do you have valid public liability insurance?** Your research institute should be able to help with this (see below).	
Risk	**Do you have a risk assessment?** You should complete a risk assessment for each of your initiatives, and have it signed off by both your research institute and any external venue (see below).	
Materials	**Do you need any materials?** Prepare any resources and take-away materials, and include extra copies of everything just in case.	
Venue	**Have you considered your AV requirements?** If you require computer/Wi-Fi access then make sure that this will be available. Where possible bring your own equipment (e.g. portable speakers) that you know will work, and have a back-up plan in case of a power/IT failure.	
	Have you confirmed the room with the venue? Contact a representative of the venue in advance to make sure that it is set up appropriately.	

Have you got adequate signposting?
Make it easy for your audience to find the venue, and that once they are there they know where the toilets, parking, and other amenities are.

Have you thought about accessibility?
Try and pick a venue that can be accessed by everyone, and work with the venue to create a safe and inclusive space for all.

DELIVERY

Participant information	**If working with potentially vulnerable audiences have you received consent from a parent/guardian and emergency contact information?** This should all be kept secure, and destroyed after the initiative.
	Have you printed off photo and video consent forms? For larger initiatives, signpost that filming/photography will be taking place and offer stickers to identify those people who do not want to be filmed or have their images used.
Health and safety	**Are you aware of the fire procedure?** Check with the venue where the fire assembly point is, and if there is any planned fire drill. Make the participants aware of this information and also ask the venue to temporarily turn off any smoke alarms if you plan on generating any smoke (e.g. through demonstrations).
	Do you know how you would access First Aid? You should have easy access to at least one person professionally trained in First Aid, with an up-to-date qualification. You might also consider asking an organisation such as St. John's Ambulance (in the UK) to provide assistance for larger audiences.

EVALUATION

Monitoring/ Evaluation	**Have you thought about feedback?** Record how many people have attended your initiative, and produce a short personal summary for reflection. Also, create a feedback survey for all participants (including other facilitators) to fill in.
	Have you done a proper evaluation? Think about what hypothesis your initiative is trying to test, and design an appropriate way of assessing this. If you are using data or information from any of the participants then make sure you have ethical clearance from your research institute, and that you have the informed consent from all participants.

(Continued)

(*Continued*)

Have you thought about…?	Check
Have you advertised your success? Consider writing a blog post (see chapter 7) about your experiences, and share any particularly engaging images via the social media accounts of your research institute and any host venue; first checking that you have permission to use any images in this way.	

Exercise: risk assessment

For any science communication initiative, you should carry out a detailed risk assessment, and any external venues (including schools) will need to have a copy of this information in advance. Ask one of the health and safety officers at your research institute for an appropriate form, or alternatively ask the venue if they have a standard format that they like to use. When completing your risk assessment, give thought to how all of the different participants (facilitators, audience members, yourself) might be at risk, and what reasonable steps you can take to mitigate that risk. For example, if you are using wired microphones, then any loose cables should be securely fastened to the floor to reduce the risk of tripping. Once you have completed your risk assessment it should be signed off by the appropriate health and safety officer at both your research institute and at any external venue that you are using.

Certain venues may also require you to have public liability insurance, and so this is a conversation that you need to have with the legal team of your research institute. Usually your initiative will be covered by your research institute's own public liability insurance (even for external venues), but they will need to be fully informed of what you are doing. Failing to do so could make you liable for any damages in the event of any accidents or injuries.

Think about the outreach initiative for the classroom that you developed in the previous exercise. Create a risk assessment for this initiative, using the suitable form(s) from your research institute. Also, take the time to find a copy of your research institute's public liability insurance, and determine if it would cover such an initiative.

5.11 Examples of science communication initiatives

Listed below are some examples of successful science communication initiatives, each of which has a well-defined objective for a specific audience, with a format that has been chosen accordingly. They serve to highlight the wide range of science communication initiatives that can be developed, and will hopefully serve as inspiration for your own.

Science Ceilidh

The Science Ceilidh [34] is an award-winning educational organisation based in Scotland that aims to bring people together with science and traditional music and arts. A ceilidh (pronounced kay-lee) is a Scottish Gaelic word for a community gathering, and traditionally it is a social initiative with folk music and singing, traditional dancing, and storytelling. The Science Ceilidh builds on this community aspect and involves participants learning about scientific research through the medium of dance and song.

The Science Ceilidh has developed a full educational and community involvement programme which has, to-date, inspired over 12 000 schoolchildren to explore the cutting-edge interdisciplinary science behind music, creativity, and learning. It has also empowered participants from across Scotland, helping rural communities to explore topics ranging from Gaelic bilingualism and health to creativity and wellbeing. It is an excellent example of how an interesting concept (communicating science through traditional music and dance) can be turned into a hugely impactful and multi-faceted science communication initiative.

Cell Block Science

Cell Block Science [35] is a unique public engagement with research partnership, which aims to promote and enable STEM in prison learning centres. It was established by the Public Engagement team at the University of St Andrews in 2016, and is currently working with the Scottish Prison Service to provide hands-on science inspiration for learners within a prison environment, bringing scientific researchers into these prisons to discuss their research and other scientific topics with the prisoners.

Funded by the Wellcome Trust, Cell Block Science began as a pilot project in a couple of prisons, during which time the capacity for meaningful engagement with an underserved community was effectively demonstrated. As well as providing STEM learning and enrichment opportunities for prisoners, Cell Block Science also creates an opportunity for scientists to interact with a public that they might not previously have considered to engage with. In conducting a detailed, and ongoing, evaluation of their initiative, Cell Block Science have also highlighted the value of science learning in prisons, sharing their best practices with a wider European network of prison educators.

Catan®: Global Warming

Catan is a multiplayer tabletop game with global sales of over 20 million copies. In *Catan*, players compete to be the first to tame the remote but rich (fictitious) island of Catan, by building roads, settlements, and cities, and in order to build these various elements, players must gather resources (brick, lumber, ore, grain, and wool). *Catan: Global Warming* [36] was designed as a 'print and play' expansion to the traditional version of *Catan*, tasking players with managing the costs associated with the use of the island's resources, and the impact of its growing settlements on the environment.

Catan: Global Warming was developed to be played without facilitation, drawing on a growing body of research that has shown how tabletop games can create a safe space for meaningful dialogue [37–39]. To date, the expansion has been downloaded and played by over 3000 people from across the world, and evaluation has demonstrated that it engenders discussion around global warming both at, and away from, the tabletop. In addition to developing this resource for a public of tabletop gaming enthusiasts, *Catan: Global Warming* has also been played by several hundred schoolchildren across the UK, where it has been used as an educational resource to support the teaching of climate change in the school curriculum.

Rhyme and Reason

Rhyme and Reason [15] was a public engagement initiative that consisted of a series of workshops run across the UK, in which people from traditionally underserved communities (refugees, asylum seekers, and people living with severe mental health needs) engaged with environmental scientists by writing poetry together. These workshops were run in the community spaces of the non-scientists, and were developed to create a platform for these audiences to discuss their thoughts and fears about topics relating to environmental change (e.g. air pollution, global warming, soil degradation, etc). This approach helped to break down the hierarchies of intellect that were discussed in section 5.3, by creating a shared space in which the non-scientists could freely discuss their opinions, and where the scientists could freely display their emotions.

By analysing the poems that were created during these sessions, researchers were able to demonstrate how this approach created a powerful way of generating what underserved audiences really know and think about environmental change, presenting a framework through which to understand differently, the lifeworld of these communities. Furthermore, bringing together scientists and non-scientists through poetry gave voice to the under-heard, giving those who could enact change an opportunity to listen.

Exercise: developing *your* initiative

As these examples demonstrate, science communication initiatives involve a wide variety of formats, many of which have been inspired by the hobbies of the scientists involved. What do you like to do in your spare time, outside of your scientific pursuits? Take a couple of minutes to think about ways in which your hobbies and pastimes could be used to discuss either your scientific research, or a more general scientific topic. Materials Science and football? Geomorphology and cakes? Planetary Science and yoga?

Linking your professional and personal lives in this manner is a powerful way to create a unique science communication experience. Choosing areas in which you have previous expertise will also give you further confidence that you have the required skills for an effective and successful delivery, and will give you direct access to a community to which you already belong.

5.12 Summary

This chapter has outlined how scientists can work with non-scientists, introducing some of the nomenclature of science communication, and differentiating between outreach (the one-way communication of ideas, from scientists to non-scientists) and public engagement (a two-way discourse between scientists and non-scientists), providing examples of each. This chapter has also provided guidance for developing your own science communication initiatives, highlighting the need to: define your objectives, consider your audience, and explore suitable formats. Furthermore, it has discussed how to fund, advertise, and evaluate your initiatives, and in doing so has asked how you might contribute towards the development of original knowledge in the field; for example, through peer-reviewed publications.

Delivering effective science communication initiatives can be a time-consuming and resource-draining task. However, it is also an extremely rewarding pursuit, which can help to further develop skills that are useful in other areas of academia (and beyond), such as supervising, presenting, and networking (see chapter 9). At times, the pressures that are placed on scientists via research, teaching, and administrative responsibilities means that it is unreasonable to expect us to also excel at developing and delivering innovative science communication initiatives. In order to ease this workload, you might also consider working with professional science communicators and social scientists. These experts will be able to ease the burden associated with logistics, assist in the setting of clear, long-term objectives, and help to effectively evaluate the process, providing of course that you involve them at the very beginning of your plans.

5.13 Further study

The further study in this chapter is designed to help you think more about developing and delivering a science communication initiative:

1. **Evaluate an initiative.** Find an upcoming science communication initiative in your local area and attend as a participant. Can you identify the objectives of the initiative? Is it aimed at a specific audience? Is the chosen format suitable and appropriate? Make a note of everything that you enjoyed and disliked about the experience, and use this to help critique your own current and future initiatives.
2. **Become a citizen of (another) science.** Find a citizen science project that is of interest to you, but which does not necessarily correlate to your current area of research. For example, if you are an astronomer then consider taking part in a national wildlife survey. When you join this project consider how much agency and ownership you are granted by the process. Do you feel like you are genuinely collaborating in the development of new knowledge, or are you nothing more than an unpaid labourer? Use this experience to help shape any future citizen science project (or other initiative) that you are developing.
3. **Get out there.** Track down the designated outreach, public engagement, or schools liaison officer at your place of work. Ask for their advice about your science communication initiatives, and ascertain if there is any funding and/ or training available to you. Making them aware of the work that you are

doing will also make you more likely to be considered for future science communication opportunities.

5.14 Suggested reading

A concise, thoughtful and easy-to-read review by D. B. Short [40] provides a brief history of science communication in the UK, focussing on initiatives following the publication of the impactful Bodmer Report in 1985 [41]. *Science Communication: A Practical Guide for Scientists* [2] provides an excellent resource for developing and delivering science communication initiatives, and includes a number of useful and inspiring case studies. Similarly, 'Delivering effective science communication: advice from a professional science communicator' [8] provides further practical advice for developing your objectives, considering your audience, and exploring new formats. Finally, the references in this chapter represent an excellent starting point for finding out more about the history of science communication, and how it continues to develop as an academic discipline.

References

[1] Bauer M W 2009 The evolution of public understanding of science—discourse and comparative evidence *Sci. Technol. Soc.* **14** 221–40

[2] Bowater L and Yeoman K 2012 *Science Communication: A Practical Guide for Scientists* (New York: Wiley)

[3] Grand A, Davies G, Holliman R and Adams A 2015 Mapping public engagement with research in a UK university *PLoS One* **10** e0121874

[4] Bucchi M and Trench B 2014 *Routledge Handbook of Public Communication of Science and Technology* (Oxford: Routledge)

[5] Brownell S E, Price J V and Steinman L 2013 Science communication to the general public: why we need to teach undergraduate and graduate students this skill as part of their formal scientific training *J. Undergr. Neurosci. Edu.* **12** E6 (PMID: 24319399)

[6] Besley J C, Dudo A and Storksdieck M 2015 Scientists' views about communication training *J. Res. Sci. Teach.* **52** 199–220

[7] Bubela T *et al* 2009 Science communication reconsidered *Nat. Biotechnol.* **27** 514

[8] Illingworth S 2017 Delivering effective science communication: advice from a professional science communicator *Seminars Cell Develop. Biol.* **70** 10–6

[9] Chilvers J and Kearnes M 2015 *Remaking Participation: Science, Environment and Emergent Publics* (Oxford: Routledge)

[10] Prokop A and Illingworth S 2016 Aiming for long-term, objective-driven science communication in the UK *F1000Research* **5** 1540

[11] Illingworth S, Redfern J, Millington S and Gray S 2015 What's in a name? exploring the nomenclature of science communication in the UK *F1000Research* **4** 409

[12] McLoughlin N *et al* 2018 *Climate Communication in Practice: How are we Engaging the UK Public on Climate Change?* (Oxford: Climate Outreach)

[13] Scheufele D A 2018 Beyond the choir? the need to understand multiple publics for science *Environ. Commun.* **12** 1123–6

[14] Illingworth S, Bell A, Capstick S, Corner A, Forster P, Leigh R, Loroño Leturiondo R, Muller C, Richardson H and Shuckburgh E 2018 Representing the majority and not the

minority: the importance of the individual in communicating climate change *Geosci. Commun. Discuss* **2018** 1–24
[15] Illingworth S and Jack K 2018 Rhyme and reason-using poetry to talk to underserved audiences about environmental change *Clim. Risk Manag.* **19** 120–9
[16] Involve: Resources https://involve.org.uk/resources (Accessed 16 October 2019)
[17] STEM Ambassadors https://stem.org.uk/stem-ambassadors/ (Accessed 16 October 2019)
[18] Bell P *et al* 2009 *Learning Science in Informal Environments: People, Places, and Pursuits* Vol. 32 (Washington, DC: National Academies Press)
[19] Gjersoe N L and Hood B 2013 Changing children's understanding of the brain: a longitudinal study of the Royal Institution Christmas Lectures as a measure of public engagement *PLoS One* **8** e80928
[20] Denson C *et al* 2015 Benefits of informal learning environments: a focused examination of STEM-based program environments *J. STEM Educ.* **16** 1
[21] Dance A 2016 Science and culture: avant-garde outreach, with science rigor *Proc. Natl Acad. Sci.* **113** 11982–3
[22] Sturgis P and Allum N 2004 Science in society: re-evaluating the deficit model of public attitudes *Public Understand. Sci.* **13** 55–74
[23] Dickson D 2005 The case for a 'deficit model' of science communication *SciDev.Net* (https://www.scidev.net/global/communication/editorials/the-case-for-a-deficit-model-of-science-communic.html) (Accessed 16 October 2019)
[24] Farache F *et al* 2019 *The Role of the Individual in Promoting Social Change, in Responsible People* (Berlin: Springer) pp 1–12
[25] Grand A 2015 Cafe scientifique *Encyclopedia of Science Education* (Berlin: Springer) pp 139–40
[26] Association B S and s.m. simple 2010 *Camous Science: The Science Busking Pack for National Science and Engineering Week* (https://worc.ac.uk/documents/CampusScience.pdf) (Accessed 16 October 2019)
[27] Lintott C J *et al* 2008 Galaxy zoo: morphologies derived from visual inspection of galaxies from the sloan digital sky survey *Mon. Not. R. Astron. Soc.* **389** 1179–89
[28] Eveleigh A *et al* 2013 'I want to be a Captain! I want to be a Captain!' Gamification in the Old Weather Citizen Science Project *Proc. of the First Int. Conf. on Gameful Design, Research, and Applications* (New York: ACM)
[29] Reges H W *et al* 2016 COCORAHS: The evolution and accomplishments of a volunteer rain gauge network *Bull. Am. Meteorol. Soc.* **97** 1831–46
[30] Illingworth S, Muller C L, Graves R and Chapman L 2014 UK Citizen Rainfall Network: a pilot study *Weather* **69** 203–7
[31] Tweddle J C *et al* 2012 *Guide to Citizen Science: Developing, Implementing and Evaluating Citizen Science to Study Biodiversity and the Environment in the UK* (NERC/Centre for Ecology & Hydrology)
[32] NCCPE: Funding http://publicengagement.ac.uk/do-engagement/funding (Accessed 16 October 2019)
[33] Google 2019 Google Forms https://google.com/forms (Accessed 16 October 2019)
[34] Science Ceilidh http://scienceceilidh.com/ (Accessed 16 October 2019)
[35] Cell Block Science https://gla.ac.uk/colleges/mvls/researchimpact/publicengagement/engagementopportunities/cellblockscience/ (Accessed 16 October 2019)
[36] Illingworth S and Wake P 2019 Developing science tabletop games: 'Catan®' and global warming *J. Sci. Commun.* **18** A04

[37] Lean J, Illingworth S and Wake P 2018 Unhappy families: using tabletop games as a technology to understand play in education *Assoc. Learn. Technol.* **26** 13
[38] Renshall H 2018 Inspiring through games *Phys. World* **31** 17
[39] Wake P and Illingworth S 2018 Games in the curriculum *Learn. Teach. Action* **13** 131–44 (https://www.celt.mmu.ac.uk/ltia/Vol13Iss1/10_Wake_Illingworth_Games_in_the_Curriculum.pdf)
[40] Short D B 2013 The public understanding of science: 30 years of the Bodmer report *School Sci. Rev.* **95** 39–44
[41] The Royal Society 1985 *The Public Understanding of Science*

IOP Publishing

Effective Science Communication (Second Edition)
A practical guide to surviving as a scientist
Sam Illingworth and Grant Allen

Chapter 6

Engaging with the mass media

There are only two forces that can carry light to all the corners of the globe ... the Sun in the heavens and the Associated Press down here.
—Mark Twain

6.1 Introduction

As scientists we are driven to explore the unknown and analyse information honestly and rigorously. This curiosity to know the truth is, and should be, the essence of our professional identity. However, it is also incumbent on us as scientists to record and communicate any original knowledge that we discover and to inspire others to be curious about nature. There is little point in being the sole person to know something and so it is our duty in equal measure to discover, disseminate, and inspire new science.

The means by which we communicate our science to stakeholders (more on who they are later) are the various 'media'. Depending on the group or individual with whom we may wish to convey knowledge, different forms of media (or methods of communication) may be relevant. In academic circles, peer-reviewed journals (chapter 2) and scientific conference presentations (chapter 4) may be the most obvious media for this relatively closed audience, but it is also important to bring our science to the attention of non-scientists, and to use science to inspire and empower the wider society (chapter 5). These latter forms of communication are often daunting to some scientists who typically (and perhaps stereotypically) feel much more comfortable communicating with their peer group. It is not unusual for researchers and academics to struggle to break down their often-technical scientific understanding for a wider audience. This potential mismatch between self, narrative, and audience can lead to all manner of misunderstandings when dealing with the 'mass media', sometimes with subtle but important consequences for the direction of public and policy debate. However, these potential misgivings should not put us off engaging with the mass media; instead we should embrace it as a powerful tool by which our science can really make a difference and have impact. After all, this is the point of our work. But it is important to know how to effectively wield this power for the purposes of truth, awareness, and meaning.

In this chapter, we focus on what we might more conventionally think of as the mass media, such as the mass communication methods of television, radio, and the printed press. The specific cases of engaging with modern social media and the Internet will be discussed in chapter 7. Here, we shall discuss how to construct a useful and succinct narrative for the often fast-paced environment of mass media, and how to remain focussed under the often stressful, and sometimes hostile, scenario of being interviewed by journalists and presenters. And finally, we shall discuss how to both maximise impact and also bring science to the attention of those that might use it to make decisions.

6.2 Why, when, and how to engage with the mass media

Mass media are routes via which science can raise awareness among a large number of people about the conclusions and implications of research in terms of what it may mean for human society and the natural world. By spreading knowledge wide and far, others may see possibilities to take your research in all sorts of tangential directions, by linking it with their specialist knowledge in disciplines that may be otherwise inaccessible to more specialist media. In other words, mass media is a vehicle to brief enormous audiences that may not otherwise seek out information from more academic settings.

Mass communication is also a means by which we can inspire the next generation of scientists, and instil the philosophy of science and the honest pursuit of knowledge as cornerstones of our civilisation and culture. A respect for the truth, and the freedom to pursue it, is undoubtedly the reason that *Homo sapiens* have been so successful on Earth to date (depending on one's definition of success). By conveying that sense of academic freedom and knowledge to others, we encourage and empower others to question, to reach informed opinions, and to rationalise and understand the world around them. We can recall childhoods filled with TV documentaries by the likes of Carl Sagan and David Attenborough, whose knowledge and inspiration drove us to study science and to think about our place in the Universe. Such luminaries are the defenders of truth and reason and we should not shy away from doing our part.

This brings us to the question of when and how we should seek to engage with the mass media. In the examples of Carl Sagan and David Attenborough, their wide and expert knowledge of whole swathes of science (as well as their innate passion, charisma, and the skills of expert producers in the background) made them ideally placed to become the icons that they undoubtedly are. However, for most of us, especially at the start of our career, we must decide when we have a story worth telling, or a comment worth making, and how best to communicate it.

Engaging with the mass media is either a passive or an active process. You may wish to alert the media to something you have to say (e.g. through a press release) or you may be consulted for comment (e.g. by a request for comment through your organisation's press office). It will usually be up to you to determine when you have something worth saying, and it is always important to ask yourself if you have valid, accurate, and useful information to convey. However, a research institute's press officers can help to suggest when there is research done by a particular scientist that is of interest to the media. Many scientists don't realise when their research may be of interest to the media, and there are also some who think their science is exciting for journalists when it isn't. Getting in contact with your institution's press office (or the press office of the publisher that you've submitted your work too), means that they can help you to find out whether your research is newsworthy, and if it is, how best to go about ensuring that it reaches the widest possible audience. The next step is to consider your narrative or message, to think about the audience you want to reach, and how best to say it in a succinct but accurate and informative way. In the rest of this section, we shall look at some of the ways you can engage with the media and how they may pick up your story.

6.3 Press releases

Press releases are an active way (from your point of view) of engaging the mass media. These are a useful tool when you know you have an important story that you feel a wider audience may want to hear.

A press release is typically a short (often one page of A4 at most) description of some new scientific conclusion, or exciting new project that carries interest to the media. It will typically contain a short title (perhaps a sentence), a description of the science and why it is important news, and often contains one or two quotes that might be used in addition to contact information for journalists to get in touch with you or your team for further comment. Drafting a good press release typically requires training or skill and it is often appropriate to get help from professionals such as a press officer (if you have access to one). However, with experience, an individual can draft a good press release with minimal help. One thing to bear in mind is that you must always seek approval for, and submit, a press release through formal channels at most institutes, especially where your affiliation may be used. This offers protection for you and may avoid embarrassing or legal issues if serious mistakes are otherwise made.

Many of the science news stories you will have seen on TV or heard on the radio will have first been picked up by a science journalist (who tends to specialise in scientific news) by reading a press release. Other press releases may be sent directly to specific news organisations' news desks. Those journalists will make a decision on whether they would like to pick up the story and then typically discuss it with their editor. They may then attempt to make contact with you to discuss the story further or they may take the information they need from your press release. A press release may have been sent to a large database that alerts potential journalists, or it may have been sent to specific journalists in a more targeted way. Wherever possible you should seek the advice and services of your organisation's press office, as they will know who to contact and may help you to draft your press release. It is also important to remember that any affiliation you may have is included in the release, especially when you are discussing work that relates to an activity carried out within your organisation.

We have both submitted several press releases through our research institute's press offices that have resulted in hundreds of live or recorded TV news items, radio interviews, and newspaper articles over several years. We have also submitted releases that have not attracted any media interest, and there are often several factors that are simply beyond your control when it comes to whether or not a press release is successful. Often the success or take-up of your press release might depend on the coverage of big news stories in the press at the time, or on editorial policy, which is why it can also be important to consider the timing of releasing your story to the world.

An example may be useful here. In 2012, Grant was the Principal Investigator for a funded project to measure the air quality around London from a specialised research aircraft [1], which involved measuring how a cloud of pollution from London was moving over areas far away from the sources of pollution within the city limits. This measurement field campaign coincided with the London 2012 Summer Olympics, and there was a lot of attention to the problems of air quality in the news, because of earlier potential athletic performance impacts during the Beijing 2008 Summer Olympics. As such, it was clear that Grant had something useful to add to the news debate at the time; he was also aware that being able to show people how we can make measurements of air quality from an aircraft would be a good way to showcase cutting edge research methods, helping people to understand how air quality impacts are felt much further away than the cities in which the pollution is originally emitted. As such, Grant issued a press release through his university's press office that described the project, which was picked up by the BBC Science Editor who then asked to join Grant and his colleagues on a research flight around London to film and interview the team, while they recorded and discussed data in real-time as it was measured. Before filming began, the editor and Grant discussed what each of them wanted to talk about and what questions would be asked. This allowed Grant to carefully plan his narrative and set constraints on what he would and would not be willing to talk about. Sadly, not all media interviews afford you the luxury of a detailed discussion on the contents of an interview in advance. Similarly, live interviews do not offer you the chance to re-record (see section 6.5), but setting your personal constraints in advance (if only in your own head) is always important for any interview, as we shall discuss in section 6.4.

Exercise: draft a press release

1. Think about your research or a topic of research that interests you. Make a list of some of the recent key findings from that discipline or from the results of your own research.
2. From this list, rank or group those findings in order of which you think may be of most interest to the public.
3. Create a maximum 10-word title that encapsulates your highest ranked finding or group of findings.
4. Create a 50-word summary or sub-title for this aspect.
5. In a further 200 words, explain the context and background to this aspect and explain why it is an important story for a mass audience.
6. Finally, give two quotes (up to 40 words each) that could be used without further permission from you and that convey a central message about this finding.
7. List who to contact for further information.
8. If you have access to a press office in your organisation, pass this press release to them for comment and advice (but make sure you tell them you don't want to release it).

6.4 Constructing a narrative for mass media

Formalising and scoping a message for mass media depends on what form of media you are dealing with, and how much space (in the case of a written article) or time (in the case of an interview) you are given to present it. However, there are some common rules to all media content to keep in mind to help ensure that your message is understood by as many people as possible. These are to:

- **Keep it simple.** Talk in non-technical language wherever possible.
- **Keep it on point.** Define and discuss a narrow scope and don't stray from this narrative, ideally identifying one key point you know you need to make.
- **Be clear.** Do not make vague statements and don't use ambiguous language.
- **Be accurate.** Make sure you have researched what you are saying and you know what you are talking about (otherwise why are you doing it?).

The most important thing you must keep in mind is that you must be careful that a journalist, reader, or viewer cannot pick and choose something you are not happy to say from your press release, article, comment, or interview. You may hear of people who have been aggrieved because they were misquoted or misunderstood in the press. In science, this is perhaps rarer than in a field like politics, where debate is often concerned with attitudes and viewpoints as they evolve. But contentious and emotive debates do surround science—take anthropogenic climate change, for example. Editorial policy may direct the context of how you might be quoted or questioned. However, much of the time any misunderstanding may be completely unintentional and due to an unbiased journalist simply not understanding what it is you are trying to tell them. Your job is therefore to minimise the risk of misunderstanding by carefully constructing any quote, article, or press release, and (where possible) taking the opportunity to first discuss your story with the press office, journalist, or producer informally so that you each have the chance to make sure there is a mutual understanding of the facts and the tone.

It is extremely rare that you will be led into a false sense of security and understanding only to be later thrown to the wolves through unexpected or off-topic lines of interviewing. This happened to Grant only once. He agreed to be interviewed live on a radio station about the impacts of volcanic ash on aircraft after a volcanic eruption in 2010. He was fielding so much media attention at the time that after a very quick telephone call with a polite producer telling him he would be interviewed about the science of volcanic ash in the atmosphere, he found himself personally accused (live) of being responsible for grounding aircraft over Europe and inconveniencing the lives of thousands. Without any chance to reply, the phone was put down and he never heard from the producer again. In this specific case, the producer may have just been looking for anyone that the radio presenter could have a one-way rant at. The station certainly weren't interested in a meaningful interview. Pretty much all Grant said was 'hello'. And they certainly didn't bother to find out if he was the right person to interview for what they wanted to talk about. In order to mitigate the risk of something like this happening to you, always do your own background check on the TV channel, radio show, presenter, or newspaper before diving in. And make a judgement about the chances of being allowed to present the message that you want to get across. If you find yourself faced with an interviewer, panel member, or audience question where your viewpoint or science may be attacked, it is always important to remain calm and objective, no matter how unnerving this may seem. It helps to remember that the more mainstream mass media in many countries is ostensibly concerned with open debate and public interest, and that open debate is best served through a rationalised discussion of facts from the viewpoint of the researcher. The approach of more politically-biased mass media and the rise of so-called 'fake news' can make such discourse very difficult, but our duty is to call this out where it is seen and passionately (but honestly) defend objectivity and fact. While heated debate and personal accusation can make for exciting reality television, for example, scientific debate is rarely convincing or useful to anyone when it strays too far from objective reasoning. In this scenario, it is more important than ever to remain focussed on a discussion of the facts as you understand them and not to be drawn into a wider discussion where you may not be qualified to speak. A calm and professional demeanour is always preferable while getting any message across. This is especially true when being asked for a personal opinion on a politically-charged subject. Scientists are trusted for their skill in objective reasoning and for their honesty. Straying too far into personal opinion is not consistent with such values unless that opinion is properly weighted in the context of scientific consensus, or grounded in one's own research.

It is important to emphasise that the vast majority of our dealings with the press have been overwhelmingly positive. Most journalists will take the time to make sure that they understand any story from your point of view and give you a chance to comment or change anything they write or present. Often, the more serious and professional media organisations may even go a step further and check that what you have said is accurate by consulting other sources, and you may even be asked to reconfirm your story. Only rarely may you be asked to speak or comment without having a chance to discuss the detail of any interaction in advance, even when preparing for live interviews. And most importantly, if you're not comfortable or confident that what you have to say will be accurately presented, you should say so and withdraw from the process.

Now that you have decided that you want to engage with the mass media, how do you go about constructing an infallible and accurate quote or story for their consumption? There are a number of common steps you should take to prepare beforehand, whether your means of engagement is live, recorded, or written. You need to break down the information you want to convey into simple and self-contained blocks and define (at least to yourself) where your story begins and where it ends, so that you don't veer off topic and end up talking about aspects for which you are not qualified to discuss professionally. And if you ever do mix personal conjecture and scientific fact, you should be very clear to point out which is which. Much like in the Q&A session of any scientific conference presentation (see chapter 4), don't try to answer a question that you don't know the answer to.

Here are some useful tips to use when preparing any content (including an interview) for the mass media:

1. Write a mock press release, whether you intend to submit it or not (see previous exercise). This is useful even if preparing a written article for a scientific magazine. It will help you to formalise your thoughts and present them in non-technical terms.
2. Try to read your press release from the point of view of a non-expert. Ask a non-expert for help if you have the time. Identify where there is scope for any confusion, such as vague statements or overly-assertive statements that are not as sufficiently balanced as they may need to be. Correct these or remove them. Or much like a literature review, make sure you understand all sides of any balanced arguments that you may need to raise.
3. Write down a single sentence that describes the one over-arching aspect, point, or conclusion that you may wish to get across. You may only get the chance to present one aspect, so make sure this simple message is front and foremost in your mind.

6.5 Television and radio interviews

Earlier in this chapter we looked at preparing a narrative for mass media in fairly general terms. Here we will talk about what it is like in practice to give TV and radio interviews. We will approach this from the point of view of someone doing this for the first time and we certainly fall well short of discussing how to present a TV or radio show; something that requires specialist training and experience, and likely a broader career aspiration.

Of all the mass media, exposing oneself to a television camera or a live microphone can be the most unnerving. Even after over 100 such interviews, both live and pre-recorded, it is still natural and perhaps useful to feel a little nervous. But it is equally important to keep calm and not panic. Different people will react differently—some of us are more confident than others—but with preparation, training, practise, and experience (and breathing), it can become easier and more rewarding. In this section, we will attempt to take some of the mystery out of the process of appearing on television and radio by citing personal experiences and offering some tips and advice. It is also worth noting that much of the advice presented in chapter 4 is also extremely useful for these situations.

As already discussed, preparation is the first step for any interview. This involves scoping out what you want, and don't want, to say and discussing the content of any interview or questioning with the journalist, producer, or presenter beforehand. In the case of live TV news or radio interviews, you will usually be contacted by a producer who will discuss and agree any interview with you over the phone well in advance. This may be several hours prior to, or even the day before any interview, and you will be invited to talk informally about the subject you will discuss on air. You will have a chance at this stage to make sure that both you and the producer know what you will and won't feel comfortable in discussing. This is a two-way preparative exercise—the producer will be looking to gauge how well informed you are, and whether you will be able to articulate your message live on air, while you need to make sure you ask any questions to put your mind at ease. You may then be invited and given a time to arrive at a studio, or told a time that a presenter and camera crew will come to you. This may then give you some additional time to prepare.

Live TV news interviews to camera can take one of three forms: a face-to-face interview with a presenter or anchor in a studio; a remote interview from a regional studio, where you typically will only hear (and not see) the presenter through an earpiece or headset; and face-to-face interviews with a presenter out in the field. The remote studio interview is perhaps the most unnerving to the uninitiated. You will typically meet with a producer or crew member in the Green Room of a studio, where you will have a final opportunity to discuss the interview before being taken to a sound-proofed room with a member of technical crew who will prepare you for camera and sound. You will then briefly talk over the microphone to a member of the Gallery, which consists of a team of directors and technicians, who will check that you can hear the studio and warn you of when you will be live with the presenter. At this stage it is important to take deep, slow breaths and to calm yourself as much as possible. You can typically hear the live sound feed at this stage, and so you should take the luxury of these few moments to listen to the news as if you were at home. If you are well prepared at this stage, thinking further about the interview can be counter-productive and only serve to add nerves. But it is important to find what works best for you.

Remember to breathe deeply during any pauses and try to be conscious of any body language or nervous fidgeting. A good way to mitigate this is to practise in front of a webcam or camera at home, and to watch out for anything that may not look professional on camera (see chapter 4 for further advice regarding managing yourself). Actions such as scratching rarely come across well; but appropriate use of hand movements, head tilt, and good eye contact with the camera can really help to emphasise your message. Body language such as this can be unnatural for some, but with careful thought and avoidance of more negative body language, it is possible to project confidence and clarity. Simple measures such as maintaining an upright and straight stance when sitting or standing can also help in this regard.

You may find your responses to the questions during the actual interview to be quite automatic, especially if you and the producer have scoped it out well in advance. Try to make sure that your key points have been made early. Answer any questions that you feel able to, don't answer those that you may not know the answer to, and make clear where personal opinion may be introduced.

Face-to-face, live studio interviews are perhaps a little more comfortable as you can see the presenter and benefit from being able to interact with their body language in a way that you cannot in remote studios. Field interviews are more comfortable still, since the field presenter (if not acting as an anchor) typically has some time to talk to you ahead of the live interview and discuss any questions with you further, which you may find naturally helps to put you at ease.

Live radio interviews are not so very different from those for television. The process and setting are broadly the same—you may be face-to-face with a presenter in a remote studio (or speaking on the phone), or out with a roving reporter. We recommend approaching radio interviews in exactly the same way as TV interviews, and when speaking to a presenter to behave exactly as you would on any TV interview, including using gesticulations or body language, which naturally help to project clear oral communication.

Some tried-and-tested tips on handling live interviews are to:
1. **Always be respectful.** Don't continue talking about a drawn out subject when the presenter has asked you to stop.
2. **Don't interrupt.** Or be interrupted…. If you are interrupted, and the interview continues, remember to come back to your key messages if they have not been made already.
3. **Demonstrate passion (pathos).** Speak clearly, loudly, confidently and with intonation.
4. **Be aware of yourself.** Be mindful of nervous body movements and actions like swaying, scratching your head, or playing with your clothes.
5. **Feel free to use gestures.** Use emphatic body language such as a head tilt and hand movements if these come naturally to you. But use these sparingly and with subtlety.
6. **Watch your posture.** Sit or stand as tall and upright as possible.
7. **Avoid filler.** Try to avoid using 'erm' or 'so' at the start of sentences. Instead, take a quiet moment to compose your answer if you need to. These 'filler' words are often used to help us formulate a response in stressful situations, but they do not present well.
8. **Be aware of your limitations.** Don't attempt to answer anything that you do not know about. Instead, answer by politely reminding the interviewer about what you are there to discuss, or better still, explain how such a question could be answered with further science if appropriate.
9. **Behave appropriately.** Remember you may be on the record (and you should ask if unsure), be mindful about not saying anything you wouldn't want to see reported or quoted and attributed to you.
10. **Practise in front of a camera yourself.** You'll be surprised how any recording device can naturally force you to behave as though there was really an audience there.

As well as the interviewer–responder setting of a live interview, recorded interviews can also include features for science documentaries or other media outputs. These settings are broadly similar to live interviews, with the exception that you may have the chance to re-record any sections you are unhappy with. In addition, the production team may have the opportunity to edit any material prior to release. Curiously, we have both found that simply knowing that there is the luxury to re-record material means that you are more likely to make verbal mistakes in pre-recorded settings, while the pressure of live interviews seems to always ensure that you get it right first time. This has especially been the case when recording one-way monologues for documentaries, and is perhaps due to the fact that the absence of someone asking specific questions means that you are often left to formulate your own thoughts, meaning that what you have to say becomes less of an automatic response and more of a voluntary choice. In such a setting it can help if you ask your presenter or crew to give you prompts. These could be written cues or verbal questions that remind you about what you have prepared in advance, thereby helping to break down any monologue into manageable sections. However, in all cases, it remains important to scope out and list the general content of what you need to say, especially if this concerns any important facts or figures that it may be important not to get wrong.

At this point, it might be useful to explain the difference between offering opinion and providing objective conclusions in any interview setting. An example may be useful here. Let us imagine that you are being interviewed about air quality in a major city and you are highlighting measurements that you have recorded and published. Let us also imagine that those data show that air quality is often quite poor in the area where you recorded your measurements, and exceeds some regulatory threshold that has been defined to constitute a risk to health. Finally, let us imagine that you are nearing the end of an interview and that you have described your measurements and that you have also (rightly) conveyed an objective opinion that there may be an impact for human health. This is a justified and appropriate objective opinion because it is based on your own published research, and in this example it is your own analysis that directly links your measurements to regulatory thresholds that define a risk to public health.

But what if the interviewer asks you 'in the light of your research, would you live in this city?' This is entirely the type of question you might be asked, as it is a question very much related to the public interest that the media serves, and on a topic for which you are perceived to be an expert. Take a moment to think about what you would say? Would you answer the question directly? Would you answer the question honestly? Think carefully, because your answer could be very powerful and influence the lives of many people if they trust you implicitly. If you said 'no', your answer may well be an honest personal opinion, but ask yourself if you would honestly encourage others to move out of the city for their own health; as this is the true basis of the question you have been asked.

A more objective answer to the true basis of the question might be to refer back to the science and suggest that your results relate to, for example, a fixed time period and location, and that someone's choice of where to live may not be based solely on their exposure to air pollution, and that it is a matter of personal choice, made up of many different factors. You may also wish to say that the science on health risks is based on large population studies, that risk at a personal level may well be different, and that further research is required to better understand the impacts on individuals. You could even go a little further and state that it is important that air quality should be improved through better policy. Such an approach steers your answer back toward the science you are discussing and away from a personal and emotional opinion, despite what the interviewer may want you to say.

These are incredibly difficult types of question to prepare for in advance but thinking about what questions may be asked and role-playing some scenarios with friends and colleagues can help to train you how to deal with them objectively. Very rarely, you may then be challenged on why you have not answered such a question directly. If that happens, you could politely reply by suggesting that it is not a decision you have to make, or refer the interviewer to the reply you have already given. But if you do choose to give an honest (but personal and emotive) answer, always be clear that this is what you have done, and be mindful of the authority and responsibility that your label as a scientist affords you.

Exercise: practise for a live interview

1. Pass your press release prepared in the earlier exercise to a friend or colleague who is willing to help you by acting as a TV news anchor and interviewer. Ask them to prepare a list of questions to ask you based on the press release, but ask them not to share this with you in advance.
2. Set up a webcam or video camera with a microphone in a quiet room where you and your interviewer can attempt to recreate a live interview experience. Focus the camera on you from a frontal aspect but with your interviewer out of view. This is because we want to simulate the pressure and attention on you (and not your interviewer).
3. Ask your friend or colleague to interview you about your chosen topic and record it. You could ask your mock interviewer to think of some particularly difficult questions, especially ones designed to elicit an emotive and/or personal response.
4. Watch the interview back, preferably with your friend or colleague, and reflect on how well your message(s) came across. Focus also on your style of delivery, confidence and clarity, and body language. Is there anything that you are unhappy with or which you could improve?
5. Repeat this as many times as you can until you feel more confident and natural in front of a camera.
6. To take this further, you could consider making this scenario a regular part of your professional life by recording a podcast or video blog about popular science in your field and uploading this to a video hosting site such as YouTube (see chapter 7 for more details).

6.6 Summary

This chapter has explored several methods of engagement with the mass media, and provided tips and advice on preparing for recorded media interviews from the viewpoint of a researcher wishing to convey a scientific message. The key to successful engagement concerns preparation, practice, and confidence. While engagement with the mass media can be unnerving, it is a powerful way to educate, to inspire others, and to affect meaningful change as a result of scientific progress.

6.7 Further study

The further study in this chapter is related to gaining experience with the mass media, it should make you think further about how best to get their attention and to promote yourself and your research in an effective way when you do:

1. **Pitch an idea.** Go to the website of a popular science magazine or TV show and look for their submissions page. Using the press release that you have developed in this chapter, along with the guidance for submissions, pitch an idea based around your current or future research. One potential source for submission is *The Conversation* [2], a not-for-profit media outlet that uses content sourced from academics and researchers.
2. **Listen to a science radio show.** Find a regular scientific radio show (e.g. 'The Life Scientific' with Professor Jim Al-Khalili on BBC Radio 4). Make a note of what you find interesting about the programme. Is there any aspect that you find unengaging? Could you imagine yourself being a contributor on that programme? If so, then how do you go about becoming one?
3. **Watch other scientists.** Look online for a recent TV interview with a scientific researcher. Do they come across well? Are they able to communicate their research in a succinct and entertaining manner? Do they engage with the other people in the studio? Try and observe if there are any other examples of good practice that you could learn from, or any bad habits that you potentially see in yourself and which should be avoided.
4. **Register your expertise.** An effective way to make yourself known to the mass media is by registering your field of expertise with a national science media organisation, such as the Science Media Centre [3] in the UK. Find out which register is most suitable for you and add yourself to the list.

6.8 Suggested reading

Chapters 1 and 2 of *The Sciences' Media Connection–Public Communication and its Repercussions* [4] are especially relevant to this chapter and discuss the impact of science and science communication in society. *Introducing Science Communication: A Practical Guide* [5] also offers some helpful advice on dealing and engaging with the mass media. For further information into the relationships between scientists and the mass media, the article 'How scientists view the public, the media and the political process' [6] presents a large study of UK and US scientists' perceptions of the media. Similarly, 'Assessing what to address in science communication' [7] examines how groups and individuals process scientific information and how this is used to develop personal and public opinion. To this end, the article also discusses how to present information in suitable ways for a given audience to obtain maximal absorption. Finally, 'The mobilization of scientists for public engagement' [8] presents a scientific study of the effectiveness and motivations of public engagement by scientists, and is part of an interesting wider special issue in this highly relevant journal that should be consulted for even greater depth on this topic if desired.

References

[1] O'Shea S J, Allen G, Fleming Z L, Bauguitte S J-B, Percival C J, Gallagher M W, Lee J, Helfter C and Nemitz E 2014 Area fluxes of carbon dioxide, methane, and carbon monoxide derived from airborne measurements around Greater London: a case study during summer 2012 *J. Geophys. Res.: Atmos.* **119** 4940–52
[2] The Conversation 2019 https://theconversation.com (Accessed 16 October 2019)
[3] Science Media Centre 2019 https://sciencemediacentre.org/
[4] Rödder S, Franzen M and Weingart P 2011 *The Sciences' Media Connection–Public Communication and its Repercussions* vol 28 (Berlin: Springer)
[5] Brake M L and Weitkamp E 2009 *Introducing Science Communication: A Practical Guide* (London: Macmillan)
[6] Besley J C and Nisbet M 2013 How scientists view the public, the media and the political process *Public Understand. Sci.* **22** 644–59
[7] de Bruin W B and Bostrom A 2013 Assessing what to address in science communication *Proc. Natl Acad. Sci.* **110** 14062–8
[8] Bauer M W and Jensen P 2011 The mobilization of scientists for public engagement *Public Understand. Sci.* **20** 3–11

IOP Publishing

Effective Science Communication (Second Edition)
A practical guide to surviving as a scientist

Sam Illingworth and Grant Allen

Chapter 7

Establishing an online presence

There was a time when people felt the Internet was another world, but now people realise it's a tool that we use in this world.

—Tim Berners-Lee

7.1 Introduction

The 21st century is a marvel of scientific invention and technological advancement, but arguably the greatest impact that any of this has had on society as a whole is the rapid development of the Internet—from a limited collection of static webpages used by a select few, to a ubiquitous entity that permeates almost every facet of our existence.

While it has its distractions and detractors, there is no denying that the Internet has helped to revolutionise the way in which science is conducted. We can now simultaneously share data and edit documents with colleagues from across the world, converse with them using video conferencing facilities, and instantly access millions of pages of peer-reviewed research. The Internet has also opened up a wealth of possibilities in a personal capacity, with people now able to share images, videos, and stories with friends and strangers at the click of a button or the touch of a screen.

With the Internet's capacity for sharing research and making instantaneous connections, it has become a professional necessity for scientists to develop and maintain a digital footprint. While this might at first seem onerous or daunting, establishing an effective online presence will help to broaden the ways in which you can both communicate and conduct science. This chapter has been written to provide useful advice on how to create a digital footprint that is catered to your needs, experiences, and expectations.

7.2 Blogs

One of the most straightforward and rewarding ways that you can start to build your digital footprint is by setting up a weblog, or 'blog'. A blog is an online collection of writing, in which you can write about the results and implications of your research, reflect on a recent field campaign, or raise awareness of an issue that you think requires attention. Alternatively, you might want to write a review of a recent publication that you have read, or about the political state of affairs of science in a specific country. Blogs don't just have to be words; you might decide instead that you want to share pictures from your research, or a time-lapsed video of a particularly impressive experiment. With so much to potentially share, first focus your message by asking yourself: what, why, and who? What do you want to say, why do you want to say it, and who do you want to say it to?

When determining what it is that you want to say, a sensible first step is to look at a selection of other science blogs to see what already exists. Many active researchers run individual blogs about their work and research, while blog networks such as Scientific American [1], and IFL Science [2] provide a platform for a range of scientists to share their stories. Reading these blogs, it quickly becomes apparent that the most successful blogs (in terms of both quality of content and readership) are those which have something new to say, and which say it in a strong and discernible voice. Much like when writing for a scientific journal (See chapter 2), there is no point in simply rewriting what has been done before. Equally there is little point in writing something that only a select few people in the world will understand.

Think carefully about your audience. Most science blogs tend to be written for a non-scientific audience, as doing so maximises their potential readership. It might be that you are aiming to reach a more scientific audience (e.g. other researchers in your field, or other scientists in general), but if this is the case then make sure you rationalise why this audience needs to be reached. If you are writing for a non-scientific audience, then give careful thought to any other commonalities that your audience may have. For example, if you are writing a blog about your research into environmental change, are you also targeting people who identify as nature lovers? Thinking careful about the exact audience that you want to target (and why you want to specifically reach them) will help you to focus your blog accordingly.

Here are five tips for writing a successful blog:

1. **Keep it short.** Aim to keep your blog posts somewhere between 400 and 600 words. There may be instances that call for a more in-depth account, but this will almost certainly result in a smaller readership.
2. **Use a pyramid structure.** Start with the key message, and then provide the context and background. If there is not enough of a hook in the first two sentences, then people will be unlikely to read any further. These two sentences are also what will tend to appear on Internet search engines, so they need to be alluring. As shown in figure 7.1, this style of writing is almost the mirror image of what you would expect to use when writing an article for a scientific journal (see chapter 2).

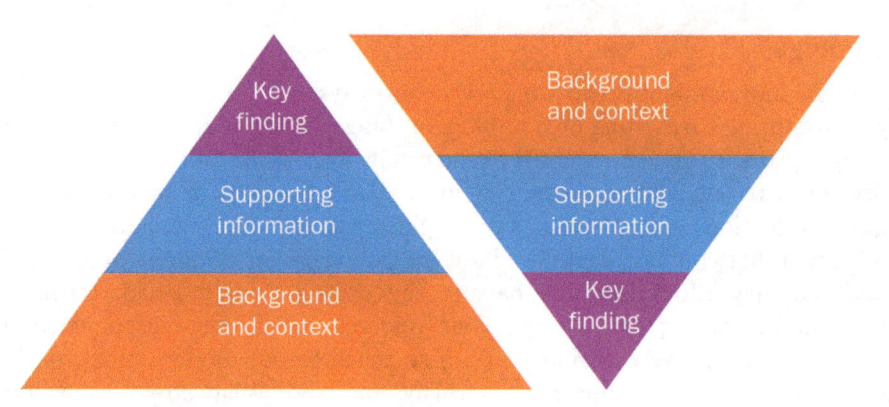

Figure 7.1. The typical structure of a blog (left) and a scientific journal article (right); the wider the section of the triangle the more content to be found in that section.

3. **Ask a non-scientist to read it.** When writing your first blog posts, get a friend or family member that doesn't have a scientific background to read your post, asking them to highlight any sections that they don't understand, or that require further explanation.
4. **Be original, and maybe even a little provocative.** Most people are looking to read something new, or a different take on something that they have seen before. Similarly, it might be more engaging for the reader if you also present your opinions or feelings (see chapter 4 for a discussion of pathos) alongside any scientific evidence, providing of course that you are willing and able to defend them should the need arise. If you do this, it is very important to be clear where science and opinion diverge.
5. **Post regularly.** Try to begin by writing one post every fortnight, increasing this to weekly (or more regular) posts once you have developed your confidence and style. People will be unlikely to keep checking your blog if you only update it every six months.

Once you have decided what you want to say, who you want to say it to, and how you are going to say it, you need to think about where you are going to host your site. There are a large number of websites that can host your blog, either for free or for an administration fee. When selecting which is best for you, spend some time reading a number of blogs across a variety of platforms to decide which most closely matches the ethos of your content. Two of the most popular blogging platforms are WordPress [3] and Tumblr [4], although there are many more for you to choose from. All of these sites provide comprehensive tutorials on the technicalities of setting up a blog, and there are also dedicated user groups within each community that can provide you with further technical support and assistance.

Interacting with other blog users, either on your blogging platform or across other blogging sites, will help you to attract followers and build a community. Similarly, if people interact with or post comments on your blog then try and respond to them in a punctual and engaging manner. Don't be afraid to defend you opinions, but as an ethical scientist you should be prepared to admit if you have gotten something wrong.

You will probably have heard about Internet trolls—people who write defacing and inflammatory comments while often hiding behind a fake identity. If you encounter any trolls on your blog (which is more likely if you are writing about a contentious topic), then as the owner of the blog you have the ultimate control. Simply delete the comment without responding to it, and report the person who sent the comment to the administrative staff of your blogging platform, if it breaches their community guidelines. The best way to deal with bullies is to starve them of the attention that they may so desperately crave.

If you find the notion of writing a blog post on a weekly basis to be a daunting task, then consider writing as part of a collective group of bloggers. Either find some colleagues with whom you share a similar vision, or reach out and interact with communities that already exist, such as ScienceBlog [5] and the PLOS Blogs Network [6]. You might also consider writing a one-off piece and hosting it on a social journalism site. Medium [7] is one of the best known examples for this type of site, being an online publishing platform that effectively acts as a blog host where writers can upload their stories. One of the major benefits of using platforms such as Medium is that unlike a personal blog you don't have to work as hard build up your readership, as you potentially have access to readers from across the entire site. However, you still need to make sure that what you write looks interesting enough so that people click on your post to find out more.

> **Exercise: write a blog**
>
> Begin by sitting down and planning out exactly what it is that you want to say, and how you want to say it. Will you be writing blog posts that detail interesting aspects of your research, or do you want to showcase some of the exotic locations that you travel to on your fieldwork? Whatever it is, try to keep the theme sufficiently broad so that you will still have something to write about in six months' time.
>
> After you have worked out the how and why of what you want to say, think about your target audience, and then go and have a look at some of the different blogging platforms. Which one works best for you? Try and plan three to five topics in advance, and follow the tips for writing a blog that are listed above; get involved with the blogging community, and respond to any comments in a timely fashion. Console yourself in the knowledge that your first few posts might not be particularly well written or well read, and that as with presenting (see chapter 4) both of these things will improve with time and practice.

7.3 Podcasts

Another way in which you can start to establish a digital footprint, or to build on the one that you already have, is by creating a podcast. A podcast is effectively an audio blog that allows you to communicate to an audience via the medium of sound. You might think that recording a podcast is a difficult and expensive process, but really all you need is a computer with some editing software (the majority of which is available for free), a decent microphone, and somewhere online to host the podcast once you have recorded it.

As with writing a blog, the first thing that you need to do is to determine what you want to say, why you want to say it, and who you want to say it to. Once you have these three things in mind the following steps should help you get started:

1. **Decide upon your format.** Will you be recording a series of one-to-one interviews, a round-table discussion, or a solo-cast? Whichever you decide upon, include some relevant noises or effects to emphasise your points or to bring your story to life. For example, if you are talking about the atmospheric effects of a recent rainstorm, then why not have some light rain playing in the background. Freesound [8] is a helpful resource for Creative Commons-licensed sounds that you can use in your podcast.
2. **Decide upon your recording and editing software.** It is worth experimenting with a few different toolkits until you find one that works best for you. Audacity® [9] comes highly recommended as a free piece of open source, cross-platform software that allows for professional recording, and which is very easy to set up and use.
3. **Find a good place to record.** If you are recording inside then make sure that you are in a quiet room that is free of noise, and where there are no possible distractions. Turn off all electronic devices that you are not using, and if you are recording using a computer, mute any email alerts etc, so that they do not interrupt the recording. If you are recording outside (e.g. at a field site), then

try to find somewhere where the background noises will lend ambience to the piece, or aid in your communications. In all instances, try to record a few seconds of background noise (i.e. with no one speaking) at the beginning and end of your recording, as this will enable you to remove any distracting sounds (e.g. the background humming of lights) when editing the podcast.

4. **Get a good microphone.** A practical USB microphone (that you can plug directly into your computer) will typically cost between £50 and £100 at the time of writing. It is well worth the investment as it will help to ensure a more professional quality to your podcast. If you do many outside recordings or interviews in different locations then think about investing in a digital voice recorder. Alternatively, buy a lapel microphone that can be attached to your smartphone, and use a voice recording app.

5. **Consider your transitions.** If the scenery change for a theatrical play is done in a heavy-handed and inconsiderate manner then it can really affect how the performance is received by the audience. The same goes for the transitions between different segments of your podcast. Carefully considered segues, as well as intro and outro music, can help to make the difference between a good podcast and a great one.

6. **Decide where you want to host your podcast.** There are many free options for you to consider, some of which have premium options if you want to upgrade to more data or take advantage of marketing opportunities. Amongst the best are SoundCloud [10] and Podbean [11].

7. **Get it listed.** Once you have found somewhere to host your podcast, you need to make sure that people can find it. In order to do this, you need to add it to a podcast directory. Arguably the biggest such directory is iTunes [12], but you should also add your podcast to Google Podcasts [13], Spotify [14], and Stitcher [15], thereby increasing the number of people that can find it.

8. **Advertise it.** Tell your friends and colleagues about your podcast, and ask them to subscribe to it and to maybe leave a favourable review on their podcast directory of choice. Use your other social media platforms to keep your followers updated when you release a new episode, and consider including a link to the podcast in the signature of your email address.

Many other of the best practices of maintaining a blog also apply to managing a successful podcast, such as posting regularly and making sure that you become an active part of the user community, rather than someone who just posts audio recordings without any further interaction.

The following five podcasts are all useful examples of the medium, each one having excellent content, good production values, and a clear target audience. They have also been chosen because they cover a wide range of formats, from panel discussions to one-to-one interviews. They can all be found in various podcast directories, and listening to at least one episode from each of them will help you to decide which approach is likely to be the most suitable for you:

1. **Why Aren't You a Doctor Yet?** A podcast that mixes serious science and tech journalism with comedy and popular culture [16].

2. **Scientists not the Science.** A podcast that explores what it means to be a scientist [17].
3. **The Infinite Monkey Cage.** A podcast that presents a witty, irreverent look at the world through scientists' eyes [18].
4. **Radiolab.** A podcast that has been designed to make a wide range of incredible science stories accessible to broad audiences [19].
5. **The Poetry of Science.** A podcast that presents new scientific research through the medium of poetry [20].

7.4 Social media

Social media can be thought of as any web-based application or website that can be utilised to communicate with a network of people. Blogs and podcasts are a form of social media, and as has been outlined above there is a large variety in the different platforms that are available within each of these spheres; for example, WordPress, Tumblr, SoundCloud, etc. In addition to these, there are also a number of other social media platforms that could and should be utilised in order to help promote, and in some cases advance, your scientific research.

New social media platforms arrive on an almost weekly basis, with the popularity of such platforms also in a constant state of flux. As such, any attempt to outline all of the different platforms would require hundreds, if not thousands, of pages, with many of the details quickly becoming redundant due to the transient nature of the interfaces. Instead, outlined below are a number of different social media platforms that are, at time of writing, amongst the most useful for developing an effective digital footprint as a scientist. An in-depth analysis of the technical details of these platforms has been purposefully avoided, as this is something that can easily change, and which is best found out via experimentation. Instead, a brief overview of each platform is given, with advice on how to best utilise that platform to maximise its potential.

When considering which of these social media platforms to use, select those that are the most suitable for what you want to achieve, and which align most closely with your own preferences in terms of interface, accessibility, and ease of use. It is very easy to become distracted by social media platforms, so make sure that you manage your time accordingly.

7.4.1 Twitter

Twitter [21] is a social media platform that enables you to connect with other users by sharing your thoughts in 280 characters or less, and when used properly it can be an extremely useful tool for effective science communication. One of these 280-character messages is called a tweet, and as well as tweeting text you can also include hyperlinks, images, and video.

The social aspect of Twitter involves 'followers'. These are the people who have decided to follow you, either because they know you or because they find what you have to say interesting or entertaining. Your tweets will appear on their Twitter timeline, just as tweets from the people who you follow will appear on yours.

If you want to tweet a specific user then you should address them using their unique Twitter handle, which is indicated by the '@' symbol. For example, if you wanted to tweet the Institute of Physics Publishing team to tell them how much you are enjoying this book, then you would write something like this:

@IOPPublishing I am really enjoying learning how to be an effective science communicator; thanks for publishing this helpful book. ☺

If you start a tweet with a Twitter handle, then the only people who can see it are those people that are following both you *and* the person that you are addressing. In the previous example, only people who were following both you *and* @IOPPublishing would be able to see your tweet. You can also use Twitter to send a direct message (or DM) to one of your followers, which will only be seen by the two of you, and is a convenient way to communicate discreetly. We advise new (and experienced) users to always check that it is a DM, and not a public tweet, that is being sent.

Here are ten tips for effective Twitter usage:

1. **Pick a good handle.** Make sure that your Twitter handle is not only unique, but that it is easy to remember; @samillingworth is a lot easier for people to recall than @poems_science_games_sam. Make sure that you also include a good profile picture and banner image, and that your Twitter bio (160 characters or less) accurately describes what you do and/or what you tweet about.
2. **Tweet regularly.** In order to gather and retain followers, you should aim to send a minimum of between three and five tweets a day. Think carefully about the messages that you want to send, and aim to send them out during the morning and evening commutes and lunch breaks, when people (in your country at least) will be most likely to be checking Twitter.
3. **Follow interesting people.** As well as following some of the most prominent names in your respective research field(s), you should consider following Tweeters who have something interesting to say about science in general.
4. **Advertise your research.** Make sure that you always tweet a link to your latest publication or talk. It will ensure that it reaches an even larger audience than it would do normally; you can find out exactly how many with the inbuilt Twitter analytics toolkit. Be sure to use Twitter to advertise all of your other digital activity as well; for example, new blog posts and podcast episodes.
5. **Use hashtags.** Using a hashtag (#) enables you to categorise a series of ideas. They also make it easier for people to find and access your tweets, as Twitter enables users to search according to hashtags (topics with the most popular hashtags are said to be 'trending'). For example, if you are tweeting about something related to science communication, try and leave room in your tweet for #scicomm, as it will help to make your tweet more visible.

6. **Be concise.** You only have 280 characters to make use of, so every letter counts. Rather than seeing this as an obstacle, you should treat it as an opportunity to hone your messages so that they are more succinct.
7. **Introduce some personality.** Your followers need to see that you are a real person, with real interests, likes, and dislikes; if there is a cause that you are passionate about then use Twitter to help promote this. Alternatively, if you do not feel comfortable combining your personal opinions and scientific research (or if your institute actively discourages such behaviour), then consider setting up both a personal and a professional account.
8. **Be polite.** Don't say anything on Twitter that you wouldn't say in a room full of crowded people, or to the face of the person you are tweeting about. This is good protocol for all social media platforms, but is especially true for Twitter, where it can be quite easy to send a tweet that you later regret.
9. **Use Tweetchats.** These involve the use of a dedicated hashtag over a set time period to discuss a particular topic. A useful Tweetchat to introduce you to the concept is #ECRchat [22], a global fortnightly discussion for the early career researcher community. It is good practice to forewarn your followers before you participate in such an event, so that they are prepared for a potential deluge of tweets.
10. **Tweet at conferences.** Use the official hashtag of the conference to increase visibility and help others to categorise your tweets. If you are unable to attend a specific conference then following the official hashtag is a useful way of finding out what is being discussed, and also gives you the opportunity to join in with any online discussions.

Exercise: compose a tweet

Think of the next presentation that you are going to give. Try and condense all of the key points into a 280-character tweet. What is at the absolute core of what you want to say, and how can you communicate this in a concise and informative manner? For bonus points, try and include some relevant hashtags.

7.4.2 Facebook

Facebook [23] is an online social networking service in which you can share information, photographs, and videos with your friends and also the wider community. By creating a profile, you are able to present your likes and dislikes, as well as aspects of your personality to everybody that you wish to share that information with. The use of security settings means that when you post certain items to your Facebook 'wall' you can decide who can and cannot see them, and posts from people that you are friends with and groups that you like will appear on your 'news feed'.

The social aspect from Facebook comes from your ability to react to and comment on other user's posts, thereby initiating a conversation that either a very small or a very large number of people can become involved with. There is also the opportunity to send personal messages, and to set up group chats, where you can also share files, much like can be done via regular email.

As well as setting up a personal profile, Facebook also presents you with the opportunity to create a page for your business or other interests, which you can then invite people to follow, and which can ultimately serve as a hub for your enterprise. Here you can share your photos and videos, create events, and invite your friends and followers to attend. There is also the opportunity for people to rate and review your page, and to share it with their friends and colleagues.

Examples of science-based Facebook Pages that have developed strong communities through a reputation for excellent content are: News from Science [24], NASA [25], Physics Girl [26], and Physics World [27]. All of these pages engage with their communities in a meaningful way, posting content that encourages interaction; for example, by posing questions, initiating dialogues, and conducting surveys. While the scope of these examples might be beyond what is achievable for an individual scientist, they serve as inspiration for how Facebook Pages can provide meaningful content to a large and varied audience.

Another way in which to utilise Facebook to engage a large audience with a particular scientific topic is through the use of Facebook Groups. While Facebook Pages are usually designed to act as the official profiles for people, brands, or businesses, Facebook Groups provide a (normally) non-affiliated platform for group communication, enabling people to share their common interests and express their opinions. There are a large number of science groups, dedicated to topics ranging from Basic Physics [28] to Quantum Mechanics [29]. Some of these groups are open (anyone can post and look at what other people have posted), whereas others require users to request access to join.

While Facebook Groups are, in theory, a space for open and constructive discussion, this is not always the case. For divisive scientific topics such as climate change, vaccinations, and evolution, discussions have the potential to veer into ugly, narrow-minded shouting matches. In an attempt to halt such intolerance, most Facebook Groups have a set of rules that must be adhered to, while others require users to answer a series of questions prior to joining, in order to try dissuade any potential troublemakers. However, for those Facebook Groups with many thousands of members, policing them is not always possible, and users might sometimes find themselves being the victims of personal abuse. In such instances, follow the same advice that was provided for dealing with Internet trolls: do not engage with the abuser(s) and report them to the Facebook Group's administrator or moderator, and also to Facebook themselves. This advice also holds true for abuse received on any other social media platform.

The responsibility that Facebook and other social media companies must take when dealing with such abuse is a topic of much debate, and is likely to be unresolved anytime soon. However, that such abuse exists at all is reflective of the society we now live in, i.e. one in which certain people believe that it is ok to abuse people because of their beliefs. As ethical scientists, we have a responsibility to callout such abuse (on all social media platforms); we also have a duty to respectfully listen to the opinions and needs of others, even if they are not necessarily the same as our own. Furthermore, we have a responsibility to actively seek out and listen to those opinions that are different from our own, as otherwise we run the risk of simply shouting our messages into a digital echo chamber (see chapter 5).

7.4.3 LinkedIn

LinkedIn [30] is a business-oriented social networking service that is mostly used for professional networking. It differs from Facebook in that it is strongly focussed on making and maintaining links with people in a mainly professional capacity. After creating a profile, which effectively acts as a digital CV, there is the opportunity to join different groups, and to connect with people that you know on either a personal or professional basis. In terms of effective science communication, LinkedIn is most effectively used as an interactive discussion board and as a job market.

If you want to use LinkedIn as an effective discussion board, then begin by selecting a number of groups that are related to your field of research and other scientific interests. Many of these groups have moderated membership, and so you will need to provide evidence about your suitability for membership, either through allegiance (e.g. university alumni groups) or through merit and/or your relevant expertise. Once you have gained membership to these groups, the discussion boards are a good way of keeping up to date with any debates that are currently happening in your field of research. Joining these discussion groups is also an effective way to connect and collaborate with other colleagues from across the world. The same advice for Facebook Groups applies here: be respectful, actively seek out other opinions, and don't just treat these groups as an advertising space for your own research and other personal projects. Try to make sure that each of your posts

Samuel M. Illingworth PhD

E: sam.illingworth@gmail.com • T: +44 (0) 161-247-1203 • Manchester, UK

uk.linkedin.com/ln/samillingworth • orcid.org/0000-0003-2551-0675

Figure 7.2. Example of a CV header, showing LinkedIn and ORCID information.

includes new material that initiates conversation, and that they add something of merit to the ongoing discussions.

In addition to being a useful resource when you are actively looking for a new job, LinkedIn also provides a shop window for future opportunities, some of which you might not even have been aware existed. By maintaining a current profile, and taking an active role in a number of groups and online discussions, you can proficiently market yourself to a wide range of future employers. In order to enhance your visibility, make sure that you keep your profile up to date with regards to your employment history, qualifications, and any awards or accolades. Creating a unique LinkedIn URL (which is free to do) is also recommended, as this can then be placed at the top of a more traditional CV, providing any potential employers with the opportunity to easily find out more about your achievements. Figure 7.2 gives an example of what the header of a typical CV might look like, complete with LinkedIn and ORCID (see section 7.5) identifications. Uploading publications to LinkedIn can be quite time-consuming, so it is advisable to upload three to five key publications that best represent you current research portfolio, alongside a link to other databases (e.g. ORCID) where interested readers can easily find all of your publications.

LinkedIn also presents you with the opportunity to list your skills, which can then be endorsed by your connections, providing evidence to future employers and collaborators that you are a recognised expert in the field. Furthermore, LinkedIn offers the opportunity for longer endorsements, in the form of 'Recommendations' from previous employers and colleagues, all of which can serve to further demonstrate your skills and expertise.

Exercise: update your LinkedIn profile

Many of you might have a LinkedIn profile, but when was the last time that you updated your information? Begin by updating your biography, choosing a profile picture that is professional in appearance, and which ideally comes from the current decade. Make sure that you provide up to date, succinct, and considered content with regards to your education, skills, and experience, then ask some current or previous collaborators to write you an endorsement in the form of a Recommendation. Finally, make sure that you have included any recent accomplishments, and apply to join some groups that are relevant to your scientific areas of interest.

7.4.4 YouTube

YouTube [31] is a video-sharing website, where users can post videos, create playlists and interact with other users by reacting to and commenting on their posts. The central hub of these activities is a user's YouTube channel, and it is where all of the videos from a certain person or organisation are grouped together. These channels can then be subscribed to, ensuring that as a viewer you are kept up to date with the most regular releases from your favourite YouTubers.

In addition to being a useful resource for cute cat videos and DIY instructions, YouTube also hosts a number of innovative science channels, which explore different facets of science and which serve as excellent examples of effective science communication in action. Two of the most widely celebrated are minutephysics [32] and SciShow [33], both of which have several million subscribers, and which provide engaging content covering a range of scientific topics, from exploring dark matter to analysing the discoveries of well-known scientists.

If you are thinking of creating your own YouTube channel, then consider all of the advice that has previously been discussed in this chapter in relation to blogs and podcasts. In addition to this, it is highly recommended that you collaborate with colleagues who have filming and visual editing experience, as there are many examples of potentially engaging YouTube videos that are let down by the amateur nature of their filming. In building your list of subscribers, engage with other YouTube communities and consider doing some guest videos with other successful YouTube vloggers (video bloggers), in order to share your work with a wider audience.

Exercise: get inspired by video

Go on to YouTube and watch a couple of videos from either the MinutePhysics or SciShow channels (or another of your choice). Then scroll down and read a selection of the comments for the most popular videos, and see if you can determine what the audience connected with. If you have something to add to the discussion then feel free to do so.

7.4.5 ResearchGate

ResearchGate [34] is a social networking site that, unlike the other examples mentioned so far, has been designed primarily for use by researchers. ResearchGate is used by researchers to share their publications, engage their peers in discussions, and search for future collaborators and job opportunities.

After building a profile that outlines your areas of research and expertise (in a similar fashion to that of LinkedIn), you are also able to upload all of your publications, either manually or by using Digital Object Identifiers (DOIs). As with LinkedIn, colleagues can endorse you for specific skills and expertise; there is also a jobs board, which recommends jobs based on your expertise and publications portfolio.

One of the most useful features of ResearchGate is that other researchers can ask you questions about your work directly. ResearchGate also presents you with the opportunity to track your publications' citation rates and to monitor the number of people who have viewed and downloaded your publications from the site. Similarly, you can follow other researchers, to ensure that you are kept up to date with their recent activities.

7.4.6 Others

The social media sites that are discussed above are by no means an exhaustive list, and as discussed in section 7.4 it is a constantly evolving landscape. The sites that have been discussed in this chapter have been chosen because of their relevance, at the time of writing, to establishing a meaningful digital footprint as a scientist. Other platforms that are worth a quick mention include Instagram [35] and Flickr [36], which in essence act as community photo depositaries. Mendeley [37] and Academia.edu [38] are alternatives to ResearchGate in the researcher-only social networking category. Reddit [39] is often referred to as the 'front page of the Internet', and is effectively a group of hundreds of thousands of message boards, in which any topic you care to think of is discussed and deliberated. Finally, Periscope [40] is a social media tool that allows you to livestream (i.e. broadcast live), giving observers the opportunity to ask live questions and to interact with you in the process. As Periscope is owned by Twitter, it also allows you to connect easily with your Twitter followers and to notify them of your Periscope broadcasts.

Another useful digital utensil is IFTT [41], a web-based service that stands for 'If This Then That', and which allows you to create your own unique automated combinations or 'recipes'; for example, tweeting a link to your blog as soon as a new article is posted. This helps to ensure that all of your outputs are linked up into a coherent stream, turning your digital footprints into an elegant waltz. Social bookmarking sites such as Diigo [42] and Mix [43] also allow you to follow other people's digital footprints, helping you to track resources, opinions, and comments, and to generally keep in touch with a number of varied and appealing communities.

7.5 Digital collaborations

As well as being a vital resource for advertising your skillset, finding information, and managing your research portfolio, the Internet provides the perfect means for truly international collaboration, via nothing more than a couple of mouse clicks or the touch of a screen. Emails have long since replaced the traditional letter or fax as our communication tool of choice, but there are many other innovative and effective ways with which we can collaborate with other scientists from across the globe.

Video conferencing represents an effective way of having group meetings, with the elimination of unnecessary travel, saving time and money while also having a positive impact on the environment. There are a number of video conferencing facilities available, both free and paid for, and as with the different social media platforms it is recommended that you experiment with several to determine which of them are the most suitable for your needs. Perhaps the most well-known examples are Skype [44] and Zoom [45], both of which offer free and paid-for versions of video conferencing facilities, with screen sharing, recording, and other tools also available.

When hosting or participating in a videoconference, always test the connections beforehand, and make sure that your colleagues have both the relevant accounts and software that are necessary for them to participate. If there are a large group of active participants, then it might be an idea to communicate via the instant messaging facilities of these conferencing suites instead, as it can be difficult if 15 or 20 people want to talk at once. As with all meetings make sure that the chair enables all voices to be heard, and stick to a pre-circulated agenda.

Document sharing facilities, such as Google Docs [46] or Dropbox [47], provide a platform for collaborating on a document or presentation, allowing you to share and co-edit documents in real time. This means that you can handily create folders for different research projects, and easily share them with other collaborators, enabling them to work on them wherever access to the Internet is available. Similarly, Slack [48] is a cloud-based piece of collaboration software that includes direct-messaging capabilities, notifications and alerts, document sharing, and group chat. As Slack also offers integration with many other services, such as Google Docs and Dropbox, it is an effective way to organise research projects. Many users prefer using Slack to email, as it is much easier to navigate and keep track of conversations than it is when attempting to navigate long email threads.

Finally, one of the most important tools to have in your digital arsenal is ORCID [49], a persistent and unique digital identifier that distinguishes you from other researchers, and which can be assigned to your publications. This is extremely useful if you have used multiple combinations of your name across your publications. This personal identification number can also be applied to research grants, helping to ensure that you always receive credit for your work. Including a link to your ORCID profile is also a means by which to reference your publications in a traditional CV, where space may otherwise be limited (see figure 7.2).

7.6 Summary

This chapter has discussed the importance of creating a manageable, informative, and attractive digital footprint, offering practical advice and guidelines on how to set up and manage successful blogs, podcasts, and social media profiles.

With so much choice it is very easy to get overwhelmed; it is simply not possible to write a number of successful blogs, run a podcast, have an active presence on every social media site, and also carry out scientific research. The most effective way of building a useful and enjoyable portfolio is to dip your digital footprint into a few of the different media, determine which of them are most suitable to your own skills and needs, and then focus on creating innovative content and meaningful communities in those that you select.

One final comment relates to the issue of personal vs. professional posts. Some employees have very strict rules with regards to what you can and cannot publish on certain social media sites, and what disclaimers you must use if you do. Read these carefully, and before you post anything to any platform ask yourself this question: 'am I both willing and able to defend these statements?' Never say anything that you would not be willing to say in person at a scientific conference, and if you are discussing preliminary results then make sure that doing so will not jeopardise any future publications for you or your colleagues.

7.7 Further study

The further study in this chapter is designed to help you think about developing your online presence and digital footprint:

1. **Record a podcast.** If you think that audio is the media for you, then follow the advice given in section 7.3 and set up your own podcast. Take the time to plan out in advance roughly what you will be talking about (be warned though, as fully scripted podcasts can sound a little unnatural), how you will market the podcast, and how you intend to develop a community around it.
2. **Join a Tweetchat.** Find a Tweetchat that you think will be interesting and relevant to you, and to which you have something to contribute. After joining in with a couple of sessions, enquire about hosting one, as normally the role of facilitator is rotated around the more active participants of the Tweetchat.
3. **Create/update your ResearchGate profile.** Fill in all of your details and upload your publications. Then, take the time to connect with some colleagues and co-authors, and endorse them for any skills that you think they possess. Read some of the general questions that have been posted about topics in your field of expertise, and see if you can provide any answers or contribute to any ongoing debates. Finally, update your ORCID profile, enabling all current and future publications to be easily traceable back to you.
4. **Build a website.** If your digital footprint is starting to feel stretched, then it might be an idea to start thinking about building a personal website. Doing so will enable you to either host or advertise all of your digital output from one easily manageable location. There are many examples of free and paid-for website builders; make use of the free-trial periods that most of these builders offer to find the one that works best for you and your audience.

7.8 Suggested reading

There are several books and other resources that aim to teach you how to build a following across all possible social media platforms and beyond. 'An Introduction to Social Media for Scientists' [50] is a helpful journal article which caters for scientists specifically, while *53 Interesting Ways to Communicate Your Research* [51] provides a number of tips for communicating your research digitally. For those of you who are interested in setting up a blog, then *Science Blogging: The Essential Guide* [52] provides a useful how-to guide for communicating scientific research and discoveries online. Finally, 'Podologues: conversations created by science podcasts' [53] is an eye-opening study into how podcasting can be used as a tool for science engagement.

References

[1] Scientific American Blogs https://blogs.scientificamerican.com/ (Accessed 16 October 2019)
[2] IFL Science https://iflscience.com/ (Accessed 16 October 2019)
[3] Wordpress https://wordpress.com/ (Accessed 16 October 2019)
[4] Tumblr https://tumblr.com/ (Accessed 16 October 2019)
[5] ScienceBlog https://scienceblog.com/ (Accessed 16 October 2019)
[6] PLOS Blogs Network https://blogs.plos.org/ (Accessed 16 October 2019)
[7] Medium https://medium.com/ (Accessed 16 October 2019)
[8] Freesound https://freesound.org/ (Accessed 16 October 2019)
[9] Audacity https://audacityteam.org/ (Accessed 16 October 2019)
[10] SoundCloud https://soundcloud.com/ (Accessed 16 October 2019)
[11] PodBean https://podbean.com/ (Accessed 16 October 2019)
[12] iTunes https://apple.com/uk/itunes/podcasts/ (Accessed 16 October 2019)
[13] Google Podcasts https://podcasts.google.com/about (Accessed 16 October 2019)
[14] Spotify for Podcasters https://podcasters.spotify.com/ (Accessed 16 October 2019)
[15] Stitcher https://stitcher.com/ (Accessed 16 October 2019)
[16] Ayoob H *et al* 2019 *Why Aren't You a Doctor Yet* (Smart Material Collective)
[17] Higgins S 2019 *Scientists not the Science*
[18] Cox B and Ince R 2019 *The Infinite Monkey Cage* (BBC)
[19] Abumrad J and Krulwich R 2019 *Radiolab* (WNYC Studios)
[20] Illingworth S 2019 *The Poetry of Science*
[21] Twitter https://twitter.com/ (Accessed 16 October 2019)
[22] #ECRchat http://ecrchat.weebly.com/ (Accessed 16 October 2019)
[23] Facebook https://facebook.com/ (Accessed 16 October 2019)
[24] News from Science Facebook Page https://facebook.com/ScienceNOW/ (Accessed 16 October 2019)
[25] NASA Facebook Page https://facebook.com/NASA/ (Accessed 16 October 2019)
[26] Physics Girl Facebook Page https://facebook.com/thephysicsgirl/ (Accessed 16 October 2019)
[27] Physics World Facebook Page https://facebook.com/physicsworld/ (Accessed 16 October 2019)
[28] Basic Physics Facebook Group https://facebook.com/groups/284686778813571/ (Accessed 16 October 2019)
[29] Quantum Mechanics & Theoretical Physics Facebook Group https://facebook.com/groups/526106304119328/ (Accessed 16 October 2019)
[30] LinkedIn https://linkedin.com/ (Accessed 16 October 2019)
[31] YouTube https://youtube.com/ (Accessed 16 October 2019)
[32] minutephysics YouTube channel https://youtube.com/user/minutephysics (Accessed 16 October 2019)
[33] SciShow YouTube channel https://youtube.com/user/scishow (Accessed 16 October 2019)
[34] ResearchGate https://researchgate.net/ (Accessed 16 October 2019)
[35] Instagram https://instagram.com/ (Accessed 16 October 2019)
[36] Flickr https://flickr.com/ (Accessed 16 October 2019)

[37] Mendeley https://mendeley.com (Accessed 16 October 2019)
[38] Academia.edu https://academia.edu/ (Accessed 16 October 2019)
[39] Reddit https://reddit.com/ (Accessed 16 October 2019)
[40] Periscope https://periscope.tv/ (Accessed 16 October 2019)
[41] IFTTT https://ifttt.com/ (Accessed 16 October 2019)
[42] Diigo https://diigo.com/ (Accessed 16 October 2019)
[43] Mix https://mix.com (Accessed 16 October 2019)
[44] Skype https://skype.com (Accessed 16 October 2019)
[45] Zoom https://zoom.us (Accessed 16 October 2019)
[46] Google Docs https://google.co.uk/docs/about (Accessed 16 October 2019)
[47] Dropbox https://dropbox.com (Accessed 16 October 2019)
[48] Slack https://slack.com (Accessed 16 October 2019)
[49] ORCID https://orcid.org/ (Accessed 16 October 2019)
[50] Bik H M and Goldstein M C 2013 An introduction to social media for scientists *PLoS Biol.* **11** e1001535
[51] Haynes A 2014 *53 Interesting Ways to Communicate Your Research* (Newmarket: Professional & Higher Partnership)
[52] Wilcox C, Brookshire B and Goldman J G 2016 *Science Blogging: The Essential Guide* (New Haven, CT: Yale University Press)
[53] Birch H and Weitkamp E 2010 Podologues: conversations created by science podcasts *New Media Soc.* **12** 889–909

IOP Publishing

Effective Science Communication (Second Edition)
A practical guide to surviving as a scientist
Sam Illingworth and Grant Allen

Chapter 8

Science and policy

Science without policy is the pursuit of knowledge. But policy without science is the pursuit of ignorance.

—Anon

8.1 Introduction

The above quote appears to imply that science can do very well on its own (thank you very much), but that policy without science is a lost cause. Another interpretation is that while science can reveal an understanding about the Universe and the human beings that inhabit some infinitesimal corner of it, that understanding is useless if it does not induce positive change for the people that are discovering it. Change is indefatigable, necessary, and inevitable. After all, the passage of time is measured and defined by discrete and measurable events. Yet how change manifests itself in our everyday lives is very much the result of guiding policy, defined for us by 'policymakers', whether they be regulators, lawmakers, governments, managers, CEOs or vice chancellors. Policy defines modern civilisation, the rule of law, and the impacts we have on our planet. Whether we always agree with it or not, policy defines our world and a greater part of our lives than we probably truly realise. However, 'good' policy, or perhaps rather 'optimal' policy, is that which is informed by knowledge, coupled with the policymakers' ability to understand this knowledge and to use it to make predictions about the potential impacts that this policy may have.

This is where science comes in. Without informed policy, we are arguably at the mercy of arbitrary or subjective guidance from groups or individuals that may not be experts, or who may be biased by independent or narrow viewpoints and vested interests. Science can provide the evidence base, the wisdom, and the predictive capacity for policymakers to make the best possible choices within the constraints of the political and socio-economic climate of the day; a climate which is itself a function of the science and policy that defines our expectations, aspirations, the way we think, and the way in which we live our lives. Yet the route that this important information takes to get into the policymakers' hands is not always as optimal as it could, and perhaps should, be.

This chapter is concerned with that pathway—the route by which science informs and influences policy. We shall explore some of the established and recognised ways that science is used in decision making in the modern world, and how you can make your scientific voice heard. Policy is often made by debate and by building a consensus, and while other voices may have opposing views, it is important that we each take a role in putting forward the best-informed, evidence-based facts and opinions to those that need, and want, to hear it. Expert guidance and opinion, especially from independent academics and scientists, is much valued by policymakers and trusted by the public. But that guidance is only useful if it is heard and received in a form that can be understood, while retaining accuracy and honesty.

8.2 How science informs policy

There are manifold direct and indirect pathways that science and knowledge is taken into the consciousness of government, lawmakers and policymakers. Indirect pathways are typically less tangible and may be subject to bias. These include personal opinion formed over time based on reading articles, watching TV documentaries, interacting with social media, and the voices of organised lobby groups with a specific agenda. Such pathways clearly have an important role and can be powerful. However, communicating science through these more passive media are dealt with in other chapters in this book. Here we will concern ourselves with more direct pathways, such as submitting parliamentary evidence and guidance for best practice.

The examples we shall present are by no means exhaustive. The pathways by which science is used in policy are many and are at times untraceable. Policies that draw on scientific evidence can apply at international, national and local levels with impacts affecting stakeholders from multi-national industries to individuals. We shall offer some general advice and some example in-roads into national policy here to help you think about how science is seen, heard, and used by policymakers. And in doing so, we will be mindful of what can be done as scientists to influence decision making more generally.

Two excellent examples of modern-world policy direction and how these have been influenced by scientific evidence, are those that are associated with the regulation of the tobacco industry, and the sometimes-conflicting priorities between slowing (or mitigating) climate change and sustainable economic growth.

In the early 20th century, smoking was not widely known to be hazardous to human health, with some doctors actually, and of course incorrectly, hailing its many health benefits. As such, the industry was not regulated in the way that it is today. However, taking the UK as an example, the route by which the tobacco industry has been constrained by policy decisions such as restrictions on advertising, restrictions on smoking indoors in public places, increases in taxation, and public health campaigns, has been (arguably) slow. This is widely observed to be due to the conflict between personal freedoms, the influence of well-funded industrial lobby groups, economic impact (with both negative and positive aspects), and the now-obvious science of negative public health impact. Over time, the accumulated evidence of health impacts and a growing consensus by health professionals has led to the policy climate we see today. Very few people on the planet are unaware of the risks; and those that choose to smoke are actively or indirectly discouraged by higher taxation, less prominent advertising, and easily-accessible public health information and advice. However, this policy success, if measured in terms of the proportion of people smoking in the western world, was absolutely the result of concerted efforts by scientists to provide unequivocal evidence to policymakers that there were significant public health impacts. These scientists provided policymakers with the information from which to make decisions about the best policies to balance health, economic, and personal freedom considerations in the face of pressure from lobby groups.

In the case of climate change policy, the debate surrounding the proportionate weight of policy (nationally and internationally) to mitigate or reduce the effects of climate change, remains far from one-sided. While an almost unanimous consensus of climate scientists makes clear that anthropogenic climate change is real and happening, a significant number of policymakers, as well as a very small proportion of scientists, claim that there is no such thing as climate change, or that any change is not induced by human activity; or else maintain that a policy response to it is not justified if it negatively impacts specific industries or national economies. Thankfully, the balance of overt opinion among policymakers and governments is that steps do need to be taken to tackle this very real problem. As such, national and international agencies and organisations such as the United Nations International Panel of Climate Change (IPCC) are carefully providing and updating the evidence base and predictive capacity in a form that is readily accessible and useful to policymakers. Furthermore, policymakers are integrated into this process through the United Nations Framework Convention on Climate Change (UNFCCC), who facilitate international and legally binding responses to climate change—perhaps most markedly through the negotiation of national greenhouse gas emissions targets. This allows policymakers to make informed, collective decisions to mitigate or reduce specific tangible risks and the causes of them, and to form domestic policy to best meet agreed targets.

The enormous but effective organisation of science for this purpose through groups of experts like those comprising the IPCC, which collates and interprets the very best peer-reviewed evidence available from across the world, clearly represents a gold standard and a concerted (and costly) effort. This bottom-up approach, by which carefully chosen experts scour the peer-reviewed literature for evidence, review it, and summarise and present it in an accessible form, makes for a system that does not simply pay attention to the loudest voice in the room (or lobby). Moreover, the IPCC reports provide a regular review that highlight remaining sources of scientific uncertainty, which then sets a forward agenda for individual scientists to respond to, seek funding to explore, and consequently better inform on. Clearly, such a global challenge demands a global response and the global participation of scientists. But the fact remains that policy decisions need to be taken on the best available evidence to hand. And without a structure like the IPCC, policymakers would be awash in a cacophony of individual research papers and individual academics, each with their own favourite climate impact and research interest, as well as vocal counter claims and agendas from lobby groups and a few maverick (to be over-kind) scientists. There is still some way to go to fully predict and understand this complex field of Earth System Science and how it will undoubtedly affect human beings going forward in a world with a rising population. To further complicate matters, in a changing environment (physical, political, and economic), it will be an evidence base that may always require regular updating. However, the pathways to policy in this arena exist and have been well organised; no policymaker can strongly argue that they are not as well informed as science (and scientists) can possibly make them.

For many other fields of science, the organisation of science-policy pathways is much less formal, and individual scientists may need to be proactive in personally bringing their evidence and outputs to the attention of policymakers. In the following section, we shall explore some of the ways in which this can be done.

8.3 What you can do to inform policy

This section gives just a few examples through which your work may influence policy. In practice there are many ways that this can happen. You may be identified by specific agencies or individuals who have heard about your work in a specific field, and as a result of this you may be asked to provide advice in the form of expert reviews of government-commissioned reports, written by civil servants or other academics and think tanks. Or you may be invited to tender for contracts to provide such reports yourself. To raise your visibility in such circles, your academic track record needs to be exemplary, while you must simultaneously seek to network through science advisory groups such as those which exist in many national science funding councils. Often, direct invitations to participate in this manner are passed through word of mouth between existing expert networks, or by recommendations from other academics who may cite your academic track record as a reason to solicit advice from you, and to draw upon your specialist expertise.

Such 'top-down' invitations would normally come some way into your academic career after you have established yourself as an expert and leader in your field. With this in mind, there are more proactive routes through which you can provide input while simultaneously raising your profile in policy circles earlier in your career. For example, national parliaments typically form specialised committees composed of, and chaired, by elected representatives who are tasked with gathering evidence for debate on policy matters of national interest prior to legislation, and who debate the impacts of legislation after they have been introduced. These committees regularly issue open calls for evidence, to which anyone can provide input. Most democratic governments, at both a local and national level, work in similar ways—they consult the public and experts for guidance to inform debate and decision making. We shall explore an example of such a consultation in the exercise below.

Exercise: find opportunities to provide evidence to policymakers

Examine the webpages of your parliament's select committees and look for open calls for evidence.

Pick one of the open (or historic) calls and follow the guidance for preparing evidence. It doesn't matter if there are no open calls relevant to your field, but it may help to look for previous calls that are.

Have a look at previous committee reports and learn about how the evidence has shaped the narrative of the report and any resulting debate. Think about how your expertise could inform that narrative and how you would best present it to these policymakers.

Should you choose to submit evidence to such a committee, it is important to approach your writing in the same way you would any academic narrative by structuring it with an introductory summary (analogous to an abstract), a body of text citing appropriate references in the context of the policy being discussed, and a conclusion, avoiding the use of technical language where possible. Evidence submissions may necessarily require you to extrapolate on your knowledge to form an opinion or a conclusion on the policy in question. As in any scientific writing, it is therefore important to be clear where you are stating your personal opinions, and what comprises the evidence informing that opinion. It may also be useful to reflect on the confidence of your conclusions and identify any knowledge gaps and what science may be required to better constrain any uncertainty.

A parliamentary committee may then reference your evidence in its report or in debate, and you may even be invited to be interviewed by that committee. This may seem daunting. However, think about the positive impact that you can then have on changing the course of debate and policy. We have discussed the impact agenda in chapter 3 and how this is a necessary part of modern science funding. The use of your science in policy and the traceability of it through citable policy pathways such as this are an incredibly important aspect of your work, and your future ability to secure funding for your research. It shows that you know how to translate science into impact, inform public and political debate, and bring about change.

Another simple way that you can provide direct input to policymakers is by registering your expertise with your national parliamentary library. Parliamentary libraries are a service to members of a parliament (or congress, assembly, etc), and their librarians do a lot more than simply run a book-lending service. Much like civil servants in government departments, parliamentary librarians provide an information connection service to elected representatives with specific questions. This may be in response to a question raised in parliament, or by a member of the public or lobby group. Parliamentary libraries often collate research briefings for parliamentarians on specific subjects of topical debate [1]. To do this, library researchers will consult published literature, including peer-reviewed academic journals, and also consult an in-house database of experts, who may be contacted for advice. You can register your expertise in your field with this database by contacting parliamentary libraries, such that you may be asked for input or advice in response to requests for information from elected officials. This may even result in you being put in direct contact with them.

Parliamentary offices offer all sorts of other services to policymakers, members of parliament, and civil servants. For example, the UK Parliamentary and Science Committee publish a quarterly magazine, 'Science in Parliament' [2], which is available to all of its members. This magazine, like many of its international analogues, openly solicits ideas from academics for articles. You can suggest a topic for an article to an editor, explaining why it is relevant and topical to policymakers and you may be invited to submit an article for publication. Note that you are not likely to receive remuneration for such work; like so much in academia it is a labour of love. However, it is an excellent way to raise your profile in policy circles as a new academic with an emerging track record.

8.4 Impact from research

As discussed throughout this book, the end result of science is typically not a research paper. Science is used by ourselves (as scientists) and by others for a variety of positive reasons and this can lead to what is known as research 'impact'. Impact can take many forms. This includes: economic gain, policy change, environmental benefit, public health benefit, technology development, public awareness of science, and innovations in pedagogy. Impact can often include many of these aspects simultaneously. In most cases, impact is a result of science communication in one format or another; and the process of generating impact from science is often referred to as the pathway to impact. This pathway may include communicating science at public events to increase public awareness, or it may focus on engaging with public, commercial, or government stakeholders to change policy, or bring about a change that results in economic, environmental, or public health benefit.

From a career perspective (see chapter 9), keeping an updated record of your research impact and impact pathways can also be extremely important in public audits of the impact of science. Many countries operate an evaluative exercise to assess this periodically, such as the Research Excellence Framework (REF) in the UK [3]. The reach and significance of the impact of science in the policy arena, as well as impacts on society, health, and the environment, are typically peer-assessed and graded in the UK and elsewhere, and used to calculate public funding settlements for public research institutions such as universities. Therefore, the impact of science matters not only in terms of its direct and tangible benefit, but also in terms of personal and institutional prestige and institutional funding. More generally, the measurable impact of science also directly justifies public investment. Many national research councils audit the impact of the research they commission, to strategically direct future funding and make an evidence-based case in public spending reviews.

When attempting to formally quantify and contextualise the impact of research for evaluation, evidence is key. Claims must be backed up and others must be able to see and follow the pathway between the outputs of research and its ultimate impact independently. This can take the form of testimonials from stakeholders that have benefitted from your research, such as company directors who have used your findings to generate profit. Or it may take the form of policy documentation that cites your research, or other evidence that is based on such outputs. It may also take the form of public surveys that track awareness of science; the awareness of problems related to plastic pollution (and changes in plastic consumption and disposal) is an excellent contemporary example. Therefore, keeping a track of the pathways your research takes toward impact, and collecting evidence of impact along the way, is a vital step in the development of both your research and profile. This may include keeping active relationships with stakeholders who are using your research to generate impact, as you may wish to ask them to evidence the impact of your research at a later time.

8.5 Summary

This chapter demonstrates some of the routes by which scientific evidence is used by policymakers and stakeholders, and provides some specific routes for how you can contribute to the process proactively as a scientist. We have introduced how the direct and indirect benefits of research are known as research impact, and how these can extend well beyond the policy sphere. Impact can be thought of as the practical societal justification for our scientific work, and demonstrating impact often justifies public and commercial funding of the work that we do.

Almost universally, the people charged with making decisions and setting policy welcome the input from experts and crave evidence to help them make informed judgments for the benefit of everyone; no one likes to make a bad decision. However, except for global 'grand challenges' or matters of high national importance and public interest, the organisation of science into policy is typically bottom-up, and relies on the proactivity of individuals or the widening of existing expert networks.

You can raise your visibility by getting involved in evidence requests and by talking to those in existing networks, while simultaneously developing your track record of expertise and underpinning research. Keep an eye out for opportunities, register and subscribe to email alerts and policy publications, and make a difference.

8.6 Further study
The further study in this chapter is designed to help you think further about developing your science policy skills:
1. **Read a policy report.** Visit the UK Parliament's Recent Select Committee publications [4], and read one of the recent reports on a topic that is related to your field of research. You will find that these documents provide an excellent summary of the topic and its policy implications, and they will give you a good idea of the kind of evidence that is most frequently submitted and referred to in such reports.
2. **Join the library.** Subscribe to the email alerts from the UK Commons Library and its research service [5]. Even if you are not based in the UK, these emails will keep you up-to-date with parliamentary debates, reports, and calls for evidence in any subject area that you wish to specify.
3. **Read a POSTnote.** The Parliamentary Office of Science and Technology (POST) provides balanced and accessible overviews [6] of research from across the biological, physical, and social sciences, and engineering and technology that are used to brief UK Members of Parliament. They provide excellent summaries of many different topics, and are worth reading to both broaden your knowledge, and highlight areas that require further evidence.

8.7 Suggested reading
The Science of Science Policy: A Handbook [7] provides an overview of the current state of the science of science policy from three angles: theoretical, empirical, and policy in practice. The authors offer perspectives from the broader social science, behavioural science, and policy communities in this evolving arena, delivering insights about the critical questions that create a demand for a science of science policy. *Using Science as Evidence in Public Policy* [8] encourages scientists to think differently about the use of scientific evidence in policy making. It investigates why scientific evidence is important to policy making and argues that an extensive body of research has yet to lead to any widely accepted explanation of what it means to use science in public policy. This book, which is also available as a free online report from The National Academies Press would be of special interest to scientists who want to see their research used in policy making, offering guidance on what is required beyond producing quality research, translating results into more understandable terms, and brokering the results through intermediaries, such as think tanks, lobbyists, and advocacy groups. Finally, *Merchants of Doubt* [9] is a general interest read and cautionary tale about how poor science can negatively influence policy debate and hinder policy trajectory; and explores in depth the tobacco and climate change examples we highlighted in section 8.2.

References

[1] Missingham R 2011 Parliamentary library and research services in the 21st century: a Delphi study *IFLA J.* **37** 52–61

[2] Science in Parliament 2019 https://scienceinparliament.org.uk/publications/science-in-parliament/ (Accessed 16 October 2019)

[3] Khazragui H and Hudson J 2014 Measuring the benefits of university research: impact and the REF in the UK *Res. Eval.* **24** 51–62

[4] Recent Select Committee Publications https://parliament.uk/business/publications/committees/recent-reports/ (Accessed 16 October 2019)

[5] House of Commons Library https://parliament.uk/commons-library (Accessed 16 October 2019)

[6] POST notes 2019 https://parliament.uk/postnotes (Accessed 16 October 2019)

[7] Fealing K 2011 *The Science of Science Policy: A Handbook* (Stanford, CA: Stanford University Press)

[8] National Research Council 2012 *Using Science as Evidence in Public Policy* (Washington, DC: National Academies Press) https://doi.org/10.17226/13460

[9] Oreskes N and Conway E M 2011 *Merchants of Doubt: How a Handful of Scientists Obscured the Truth on Issues from Tobacco Smoke to Global Warming* (London: Bloomsbury)

IOP Publishing

Effective Science Communication (Second Edition)
A practical guide to surviving as a scientist

Sam Illingworth and Grant Allen

Chapter 9

Other essential research skills

People from different backgrounds approach a subject in different ways and ask different questions.
—Jocelyn Bell Burnell

9.1 Introduction

There is much more to science, and being a scientist, than writing grant proposals, performing experiments, and presenting your findings. A successful research scientist also develops and uses a range of more general skills. Some of these skills are implicitly developed through undergraduate and postgraduate learning and experience, while others are honed lifelong through continuous practice and occasional formalised training. These often invisible but highly important skills include aspects such as time management, networking, academic integrity, and self-reflection. The career pathway of the modern scientist often involves multi-tasking and dynamic adaptation to workload; this is often a source of stress and anxiety, and the best response to this can be a very personal experience. Here, we offer some top tips on how to manage your professional life based on our own experience and those of others. This chapter is by no means exhaustive but we will outline some of these important skills, and discuss why they are important for you to consider in both your current and future career.

The Royal Society's 2010 policy document 'The scientific century: securing our future prosperity' [1] concluded that the majority of people who undertake a PhD will end up pursuing a career outside of academia. It is therefore necessary to consider and develop skills that will help you to succeed in these transferable environments. Pursuing a career outside of academia does not mean that you have 'failed' or 'turned your back on science'. Rather, there are many careers outside of academia that are still connected to science, and which may well be better paid, or have more favourable working conditions suited to your preferred professional lifestyle. In our experience, it is very common that postgraduates and early career researchers become blinkered to the opportunities that exist beyond academia. They may often become laser-focussed on their niche research field, and forget that the transferable skills and qualifications gained earlier in their career are still as valued as ever by alternative employers and industry, and that such training opens many doors.

Being a scientist means that you have a number of key transferable skills that make you a genuine asset as a potential employee, or as a self-employed practitioner. However, you must learn how to recognise and advertise these skills effectively, taking advantage of any opportunities to develop them further. This can include keeping track of any events, activities, and training programmes that you participate in, as these will serve as useful exemplars when regularly updating your CV, conducting personal development reviews, or making a case for promotion, for example. Digital tools such as Vitae's 'Research Development Framework' planner [2] offer a convenient way of storing all of this information in one place, and also for identifying areas in which further skills development may be required.

In addition to developing skills that make us more effective scientists, we should also consider how we can become more ethical and apply high standards of academic rigour and integrity. We are part of a long line of practitioners, and as such we have an obligation to respect our scientific heritage, recognise and correct our mistakes, and create an inclusive environment for others. This chapter also contains advice for how we might best achieve this.

9.2 Time management

Procrastination and prioritisation are some of the biggest hindrances that we face as scientists, whether via the obvious and immediate temptations of social media, or the subtler distractions of spending too much time pursuing a project that may be of no long-term benefit. However, there are a number of basic actions that you can take in order to maximise your time efficiently:

1. **Know *when* you work best.** Every person is unique, and research has shown that different people work best at different times of day, and with variable concentration spans [3]. Determine when in the day you are at your most effective, and choose this time to focus on your most urgent and important tasks (see figure 9.2). For example, if you know that you are at your most productive at the start of your working day, then ignore the temptation to check and respond to every email and instead finish your journal article that has a looming deadline. Equally, making effective use of break times can help to reset your concentration and ultimately lead to a more productive day. When it comes to productivity, quality of time spent working is often more valuable than the quantity of time spent.

2. **Know *where* you work best.** Select the correct environment for the task that you are doing. For example, if you work in a busy or shared office, then this environment might be extremely conducive for discussing ideas for a future research project. However, reading journal articles might be better suited to a quieter room, such as at home or in a library.

3. **Avoid unnecessary meetings.** The modern workplace places many demands on people's time. Make sure that any meetings that you organise are absolutely essential, and that they are planned effectively. Only invite those people that are needed and consider structuring meetings for people to be able to attend in part (i.e. only for the parts that are relevant for them). For potentially unavoidable meetings that clash with other commitments, try to obtain the agenda beforehand and accomplish the tasks that are being discussed. Doing so will help to justify why you might not be needed at the meeting, and will help you to prioritise.

4. **Learn to say no.** If you take on too many things, then you run the risk of doing all of them badly, or you may create stress and anxiety for yourself and those that may rely on you. It is perfectly ok to say no to people, or to negotiate the best way to manage activities, and sometimes it is necessary for you to be a little selfish, to know your value, and to ask yourself if it is really worth it. If you do turn down an invitation then make it clear that you are available for future consideration (but again, only if it will be of benefit). Regularly discussing and reviewing your workload with your line manager and mentors (see section 9.6) can be important to ensure that you do not take on too much, and that those who can help you prioritise your workload know what you are doing and what is being asked of you.

5. **Manage your calendar.** Try to include daily tasks and deadlines in your calendar, including dates for follow-up and evaluation where necessary. Blocking out specific days for research or development activities can also help to avoid them being taken up by too many meetings.

6. **Manage your email.** Emails remain a widely-used default method of communication in professional workplaces to convey important (and far less important) information. Many people become anxious about achieving 'inbox

zero' and struggle to switch off until they have read and responded to every message someone else has chosen to send them. Some may find themselves interrupted from other tasks hundreds of times a day to achieve this. A simple strategy may be to only monitor emails at key points in the day; for example, at the start and end of each day (and perhaps during lunchtime). And while many value the flexibility of being able to send and receive emails outside of working hours, reflect on whether this is useful for you personally and balance this with any impact on your wellbeing (positive or negative). Equally, think carefully about the efficiency of the emails you send and think about whether email is the best medium to convey the information you might need to communicate. For example, would picking up a phone be a more efficient way to get to the bottom of a complex discussion?

The STING acronym (figure 9.1) provides a useful aide when thinking about how to manage your time effectively. Begin by selecting an appropriate task, and plan the amount of time you will need to complete it; for example: 'In the next two hours I am going to write 500 words of the introduction to this journal article.' While you are doing this task, ignore everything else (put your phone on silent and deactivate your email if necessary), allowing yourself only comfort breaks until it is finished. Once the selected task is finished, consider giving yourself a reward. This can be anything you like, from a slice of cake to allowing yourself to check your emails. When selecting the task itself, choose something that is substantial, yet ultimately achievable within a sensible timeframe.

An alternative time management strategy is the Pomodoro Technique®. This is a time management system that breaks the working day into 25 min intervals, separated by 5 min breaks. Each of these intervals are referred to as a Pomodoro, and after four such intervals, a longer break of about 15–20 min is taken. This technique has been shown to instil in the user a sense of urgency, with forced breaks helping to avoid feelings of burnout [4]. You can keep track of these intervals by using a stopwatch, or by downloading a dedicated application for your computer or smartphone.

One final time management technique is the use of an importance-urgency matrix, such as that shown in figure 9.2. If you have a number of important tasks to accomplish, then determine where they lie on this matrix. Those in Q1 need to be dealt with

Select a task

Time yourself

Ignore everything else

No breaks

Give yourself a reward

Figure 9.1. The STING acronym for time management. This provides one useful methodology for managing your work effectively.

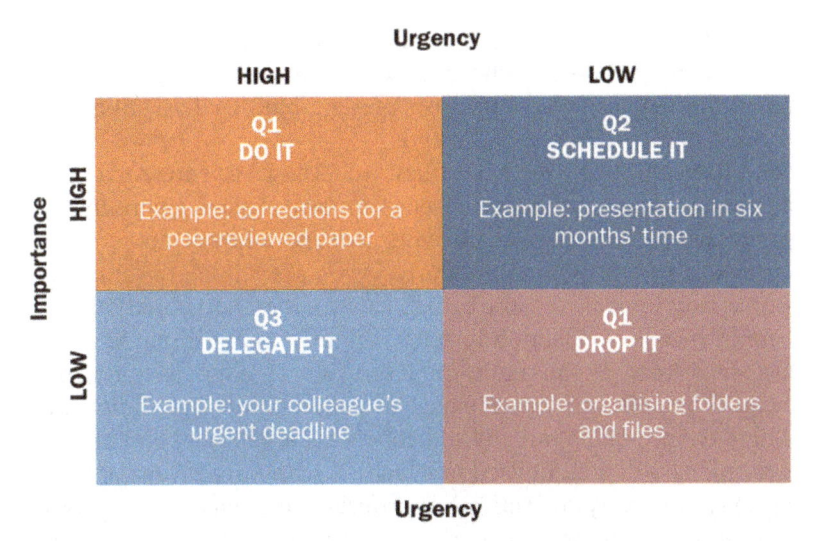

Figure 9.2. The importance-urgency matrix. This can be used to help prioritise which tasks need doing quickly, which can be postponed or delegated, and which can be dropped altogether.

immediately, followed by those in Q2, while those in Q3 might either be delegated or pushed back, and those in Q4 can probably be dropped or ignored altogether.

9.3 Networking

Networking is a skill that for many of us does not come easily. Very few would call themselves an expert networker, and some may never find it comfortable, choosing to instead develop alternative approaches to engaging with other professionals. However, as with presenting and writing, effective networking can be developed with time and practice. As a scientific researcher there are typically plenty of opportunities to network, be it either informally during coffee breaks at conferences, or in a more formal setting such as an organised dinner or dedicated networking session. In almost all of these circumstances, the biggest barrier to overcome is the initial nervousness associated in approaching a stranger and starting a conversation. The following advice should help you to overcome these nerves, and to build your confidence when a networking opportunity presents itself:

1. **Don't be afraid.** Many early career researchers struggle to engage with more senior scientists, afraid that they are too 'important' for them to talk to. However, all eminent scientists were once early career researchers themselves, and most of them will welcome the opportunity to speak to other eager and passionate researchers.
2. **Be yourself.** All of us get nervous at times, and this is even more pronounced when we are trying to be someone, or something, that we may feel that we are not. Imposter syndrome (a feeling of inadequacy) is especially prevalent in the academic community, for people of all ages and backgrounds [5]; even the most successful scientists often question their authenticity. Recognising and accepting this are the first steps in mitigating its more negative impacts.

Maintain your integrity and be safe in the knowledge that you are no doubt a very interesting person, who is an expert in their respective field(s). There may be more experienced scientists present, but this does not make your own research or opinions any less valid.

3. **Don't hog conversations.** Oftentimes, well-known scientists may have a queue of people waiting to talk to them in busy social environments such as conferences. If this happens to you, then go and talk to someone else and come back to others later. Similarly, if you are talking to someone and other people may appear to want to talk to you, then you could attempt to bring them into the conversation to allow things to move along in a natural way.

4. **Just stand there.** If you find it difficult to start a conversation then look for a group of people who are engaged in conversation and stand next to them, joining the group. Eventually someone will either start speaking to you, or an opportunity will present itself in which you can introduce yourself. Of course, this may not always be entirely appropriate and may feel very awkward, but if the group you approach are clearly discussing a private matter, they will be sure to tell you politely.

5. **Try not to be too blunt.** Networking sessions can be an excellent opportunity for seeking out potential employment for early career scientists. However, a slightly tactful approach in which you demonstrate your skill set and expertise, before casually mentioning that your contract is coming up for renewal, can be preferable to asking someone if they can employ you before you have even been properly introduced.

6. **Always carry business cards.** Doing so will enable you to continue any conversations at a later date, and will mean that your details can also be passed on to other colleagues.

If you find yourself overwhelmed at a networking event, then ask someone for an introduction. For example, if you are joining a new team or working group, or want to speak to someone in particular, then try asking one of your colleagues or even your supervisor for an introduction. This can help to remove some of the nervous apprehension from networking. Similarly, on occasions when you know that two people's work and interests would be well aligned, take it upon yourself to make the relevant introductions.

If you tend not to feel comfortable in large, social settings, you could hone your networking skills in small-group or informal networking events first. It may also help to start by going to events where you are more likely to find like-minded people (e.g. a meeting of cat-loving particle physicists), or to go to events with colleagues you feel at ease with. However, if you do end up attending a networking event with some friends or colleagues, try to avoid talking only to them, as that somewhat defeats the purpose of attending such an event in the first instance.

9.4 Teamwork

Working in a team, whether as part of a large international consortium or as a member of a small local group, is often a part of any scientific researcher's day-to-day activities. Effective teamwork requires a variety of roles to be filled by different members of the team, with each role and team member to be treated with unbiased and non-prejudiced respect. Despite the claims of several behavioural and personality tests, the best way to determine which role you are most effective in (and which you enjoy the most) is through trial and error. It might be that you are the kind of person who likes to organise, but who struggles to come up with innovative ideas. Similarly, you might be the kind of person who is excellent at seeing the bigger picture, but who sometimes has difficulty in recording those ideas in an accessible and informative manner that is essential for grant applications, etc. Your favoured or most effective role might also change depending on the project or team; don't be afraid to try new roles in new situations.

Whatever your role in the various teams you are a part of, it is typically impossible for you to be able to do everything by yourself and to a high standard, while also maintaining a healthy work-life balance. Furthermore, the days when review panels looked more favourably on solo-authored publications and lone grant applications are now thankfully behind us. Instead, internationalisation and collaboration are viewed as the key to being a successful scientific researcher. In addition to developing such collaborations you should also learn how to contribute to them in an inclusive and considerate manner.

The key to successful teamwork is in appreciating that everyone is different. This might seem like an extremely obvious statement, but the majority of disagreements in teams occur because people either assume that everyone will behave in the same manner as themselves, or else they expect them to do so. Each team member will have a different set of strengths and weaknesses in different contexts, and what may work for one person might not work for someone else. For example, if you are the kind of person who leaves everything to the last minute, but always gets it done, be sensitive to the fact that other members of your team may have prepared their contributions weeks, or possibly months, in advance and may find deadlines stressful. Equally, if you work towards completing deadlines as quickly as possible then you have to accept that other people might not be able or willing to do this, so don't harass them because they are not working to your timescale. To manage and mitigate this in complex projects, effective planning, regular review, and flexibility should be built-in to project management and design from the beginning of any project to its end. As with any relationship, working as part of a team is all about compromise and respect, and if you remain professional, committed, and polite then you will find working as part of a team to be a more enjoyable and rewarding experience.

A final, critical, component to working as part of an effective team is diversity.

Diversity is, however, not just a box to be ticked; several studies have shown that an increase in diversity leads to an increase in productivity, innovation, and impact [6–10]. Ensuring that teams are made up of a diverse collection of people with different approaches and backgrounds will mean that there is a diverse range of opinions, needs, experiences, and solutions. If your current or future collaboration is missing this diversity then ask yourself why, and then address it. If you find yourself in a position where you cannot enact change then find someone who can and raise it with them. Part of our responsibility as ethical scientists is in helping to ensure that science is for everybody, and a vital step in achieving this is in re-normalising who a scientist is and what they look like.

9.5 Objective reflection

Reflection is a fundamental part of the learning cycle—a necessary step in the development and reinforcement of knowledge and a check and balance on the accuracy of what we ultimately communicate in the form of scientific outputs. As scientists we are constantly taught to reflect upon our scientific work. For example, we may perform and repeat an experiment and then adjust certain variables based on the initial results. This reflection is a critical part of the scientific process, yet how often do we take the time to formally reflect on our scientific careers more generally?

The value of reflective learning extends far beyond analysing the results of your latest experiment; it is a practice that will help shape your research and evaluate your career path. For example, reflecting on a recent success will increase the likelihood of it being more than a one-off, while reflecting on a failure will enable you to better understand what happened and to avoid repeating similar mistakes in the future.

There are several models that can be used to help guide and structure a useful reflection; one of these, Gibbs' Reflective Cycle [11], is shown in Figure 9.3. Gibbs' Reflective Cycle is centred around the concept that reflection takes place after an experience; it provides a framework of cue questions, offering a checklist for learners as they progress through the cycle. This reflective cycle focuses on learning from experiences by involving feelings, thoughts, and recommendations for future actions. For scientists who have often been taught to ignore their emotions (see chapter 4) this can be difficult, but doing so will facilitate the creation of more

Figure 9.3. Gibbs' Reflective Cycle. This cycle features a series of questions that help to guide the user through a process of meaningful reflection.

effective future plans. If you find that Gibbs' Reflective Cycle does not work for you, then there are several other models, such as Johns' Model of Structured Reflection [12] and Jay and Johnson's Typology of Reflective Practice [13], that you can investigate; experiment with several of them until you find the one that is most suitable for you.

9.6 Mentoring

Working with a mentor will provide you with valuable advice from a more experienced person, improving your knowledge and skills, and building your professional network in the process. Many research institutes offer formal mentoring schemes, especially to early career researchers and new members of staff; such schemes can also lead to a better understanding of your organisation and its various bureaucracies. However, with more formal mentoring schemes there remains a risk that you may be paired with an unsuitable mentor, someone with whom you either have very little in common or whose networks and experiences are not necessarily aligned to your future career trajectory. In order to counteract this, and in instances where no formal mentoring scheme exists, it is advisable to establish your own independent network of informal mentors.

In choosing your mentors pick people whom you respect, and whose experience and/or networks will help you in your development as a scientist. Such a network need not be made up solely of colleagues from your immediate place of work. They could also be people you can meet up with on an irregular basis (either virtually or in person), and in whose presence you feel comfortable exchanging knowledge and advice. Your prospective mentor(s) should be someone that you get on with, and with whom you share a mutual level of respect and understanding, but they needn't be in a position that is senior to your own. Rather, they should possess a level of expertise in an area that you have identified as lacking in your own skill set. For example, if you find presenting work to a non-scientific audience difficult, then who might you know that excels at this? If you are inexperienced in writing grants then do you know someone who has recently been successful in their own application? Building up this informal network of mentors will also build your self-awareness and confidence, and could potentially lead to future opportunities to collaborate.

As well as being a mentee, you should also seek out opportunities to act as a mentor. This process can be done either formally or informally, and by exchanging knowledge with other scientists you will help to reinforce your own understanding, build networks, and gain new perspectives and fresh ideas. Furthermore, as an ethical scientist you will be helping others to determine the best path forward in their own scientific pursuits.

9.7 Career planning

As mentioned in the introduction to this chapter, the majority of people who undertake PhD study will end up in a career outside of academia. Given the limited availability of government funding, and the increasing numbers of research students who see a PhD as a gateway to employment, this seems to be a trend that will only increase in the years to come. In fact, it is probably more accurate to say that academia is the alternative career path. Pursuing a career outside of academia can be a rich and rewarding experience. However, it is necessary to have a clear plan of what you want to do, why you want to do it, and how you intend to achieve this.

If you enjoy doing scientific research, then there are plenty of careers that you can pursue outside of academia. For example, you could work for a large non-university research institute or government agency such as the UK's Environment Agency, or the Max Planck Society in Germany. Alternatively, you might want to work for one of the global tech giants, or at an instrument manufacturer or manufacturing company. Many of these jobs will allow you to conduct scientific research, publish journal articles, and attend conferences, with the added bonus of a full-time contract and the sense of security that is often absent from many non-tenured positions within academia.

If you find yourself no longer interested in scientific research at any point in your career, then there are always plenty of options for you to pursue. However, you need to think carefully about how best to market your unique skill set to a non-scientific audience. For example, writing a thesis demonstrates that you have excellent written communication and time management skills, while analysing data and setting up experiments exemplifies your problem solving skills. Presenting your research at a conference typifies your outstanding oral communication skills, while supervising undergraduates expresses your aptitude for teamwork and leadership, and doing all of the above is testament to multitasking capabilities. Recognising your experiences as evidence of the skills others may be seeking is key to developing an engaging and effective CV.

There are many jobs outside of scientific research that would benefit greatly from your transferable skills. It is just a question of finding them and not being too narrow-minded and blinkered by the minutiae of what you may have worked on in the past. A good place to start is The Versatile PhD website [14], which provides a list of non-academic careers and the potential routes into them. If you have an aptitude for communicating your research to a varied audience, then you could consider a career in teaching. In the UK (and many other parts of the world), there is a large shortage of qualified science teachers, particularly those with expertise in physics, maths, chemistry, and computing. In order to address this shortage, schemes such as that set up by the UK Government's Department for Education offer bursaries and financial support for teacher training.

If you do decide to pursue a career in academia as a research-focussed academic, then you may need to be realistic. In most countries the number of PhD students is increasing at a rate that is greater than the increase in government spending on research, or the rate at which undergraduate numbers may grow demand for teaching. This means that there are simply not enough permanent academic positions for every PhD graduate, and many excellent researchers are forced to find employment via a series of fixed-term contracts that may offer less job security and might involve relocating over long distances. To give yourself the best opportunity of achieving tenure you need to have a CV and expertise that demonstrates leadership and independence. Applying for fellowships, such as those described and discussed in chapter 3 are often a springboard into academic tenure. You also need to be open to a variety of possibilities (including moving abroad), resilient, and reflective in your approach (see section 9.5). Getting a tenured position in academia is not impossible, but it is certainly more difficult than it has ever been.

When considering your next career move, be aware that leaving academia does not permanently close the door to a future return. If an opportunity away from academia presents itself then don't dismiss it without reflection; many successful academics have spent time away from academia, and their careers have improved significantly because of this.

> **Exercise: write a five-year plan**
>
> Having a five-year plan will help you to focus your future career objectives. Taking the time now to plan out what you want to achieve over the next five years will also help to ensure that you maximise your opportunities, and will reveal which skills you need to develop further and where best to focus your time and energy. Think about what grants or fellowships you wish to apply for, how many publications you aim to produce, and any awards or accolades that you would like to receive.
>
> After writing your five-year plan, ask one of your mentors to review your initial thoughts, and to see if you are being realistic. After another iteration, start to break down your plan into milestones and achievable tasks, and then use this as a guide to help focus your work into achieving your aims. Reflect on the five-year plan during regular intervals (e.g. every six months), updating it with every major achievement, accomplishment, and setback, and what you have learnt from these experiences.

9.8 Open access

Access to knowledge is a basic human right. Yet sadly as scientists we are often forced to operate in a framework in which this is not facilitated as easily as we might wish. If you are reading this as a scientist at the outset of their scientific career, then you may be surprised to find out that it can cost (often very large sums of) money for others (especially non-scientists) to access and read the latest scientific research. Even if these fees are not being charged to you personally, the chances are that it is costing your research institution or library tens of thousands of pounds/euros/dollars that could otherwise be spent on research, resources, staff, or infrastructure.

From a European perspective, the '2019 Big Deals Survey Report' [15] provides a mapping of major scholarly publishing contracts in Europe. By gathering data from 31 consortia with five major publishers (Elsevier, Springer Nature, Taylor & Francis, Wiley, and the American Chemical Society) this report found that the total costs of the participating consortia exceed €1 billion for periodicals, databases, e-books, and other resources, almost all of which led to significant profits for these large commercial scholarly publishers.

Over the last 30 years, traditional journal prices have increased at a much faster rate than inflation, often resulting in significant profits for publishers [16, 17]. In the past, scientific journals existed for two reasons: as an affordable option for scientists to publish their work (as opposed to the more expensive option of personally-published books), and as a place where both scientists and non-scientists could find out about advances in science. Sadly, in recent times many journals seem to have lost sight of their role in providing a service to the scientific and non-scientific publics, hence the current drive for open access (OA). Open access represents a model where published research is free at the point of use for public readership, but where publication fees might be paid by the author's research institute or other funding body. We describe the different and emerging OA routes below.

The beginning of the modern OA movement can be traced back to the 4th July 1971, when Michael Hart launched Project Gutenberg [18], a volunteer-led effort to digitize and archive cultural works for free. However, it wasn't until 1989 (and with the advent of the Internet) that the first digital-only, free journals were launched; amongst them *Psycoloquy*, edited by Stevan Harnad and *The Public-Access Computer Systems Review*, edited by Charles W. Bailey Jr. Since then, the OA movement has grown considerably, although publishing articles so that they are free to access is itself not without expense. Despite the lack of print and mailing costs for fully online journals, there are still large infrastructure and staffing overheads that need to be taken into consideration, and so rather than make the reader pay, alternatives have had to be found.

One alternative, known as the 'Gold Route' to OA, is to make the author(s) of an article pay for the right to have their research accessible and free to all readers. Many journals already require an article processing charge (APC) to be paid before publication. Fully OA journals build this cost into their APC, while other journals introduce an additional charge if the author wants to make their publication free to read.

The other main alternative is the 'Green Route' to OA, which involves the author placing their journal in a central repository that is free to access. The journal in which the article was originally published will usually enforce an embargo period of a number of months or years that must pass before the published articles can be placed in these repositories, although this can often be circumvented by uploading final 'accepted for publication' drafts of the journal article to public repositories. Most research institutes will have their own repositories in which such articles can be stored, and their data management team will be able to advise you with regard to the legalities of storing and hosting your publications (i.e. what version to use, how long to wait following publication, etc).

Both of these approaches to OA have their respective advantages and disadvantages, and normally research institutions and/or funding bodies guide the route that researchers choose. For example, the United Kingdom Research and Innovation (UKRI), has a policy that supports both the Gold and the Green routes to OA, though it has a preference for immediate and no-cost public access with the maximum opportunity for reuse. Another key aim of the OA movement is that published research is free to reuse in future studies. This might seem like a fairly trivial point, but for most articles published in closed access journals, express permission is technically needed from the publishers if the results are to be used in any future studies.

The major barrier that still needs to be overcome with regards to OA is determining who pays for the rights to free access. Many governments and/or funding bodies have a centralised amount of funding, which they allocate to research institutes. However, issues arise when one considers the limitations that this imposes on poorer countries, institutes, research disciplines, and independent researchers. It is for these reasons that many organisations and individuals are investigating and developing 'OA 2.0', an initiative in which articles are: free to read, free to download, and free to publish. However, such an approach will require a major change in the funding model of almost all publishing companies, and must be cautious to retain rigorous processes in peer review, impartial editorial decision making, and legacy archiving.

9.9 Integrity and malpractice

In their 2015 report 'Seven reasons to care about integrity in research' [19, p 1], Science Europe observed the following:

> 'Research integrity is intrinsic to research activity and excellence. It is at the core of research itself. It is a basis for researchers to trust each other as well as the research record, and, equally importantly, it is the basis of society's trust in research evidence and expertise. Research misconduct is not a victimless crime and can damage reputations, careers, patients and the public. It is also a waste of public investment in research and is costly to remediate.'

Without integrity there is no meaningful research, and thus there is no science. Even with the checks and balances of peer-review systems, ethics boards, and academic scrutiny, so much of our research remains reliant upon science being conducted in a fair, honest, and transparent manner.

At this point, it is important to differentiate between 'integrity' and 'malpractice'.

Integrity represents both the ethics and the quality of academic rigour and best practice in the presentation and approach of science. This can include best practice in the conduct of experiments and experimental techniques, as well as the proper use of statistical and analytical approaches, thorough transparency of methods, and data availability. Integrity can therefore be summarised as representing standards of best practice and convention, which ensure transparency, scrutiny, and analytical repeatability. The checks and balances of science work hard to push up these standards constructively when applied well. Poor standards of integrity can therefore represent sloppy practice and ultimately useless scientific outputs, but not necessarily wilful neglect. It also includes instances of less serious plagiarism, such as careless oversight resulting in a lack of proper citation when discussing other people's work.

On the other hand, malpractice represents wilful neglect, data fabrication, wilful plagiarism, and potentially illegal and wildly unethical approaches. Thankfully instances of malpractice remain extremely rare, but they can have a very high impact and be extremely dangerous to the scientific community and society at large. Some contemporary examples of fields in which malpractice must be carefully guarded against for public good include (but are not limited to): germline gene editing [20] and the application of artificial intelligence [21]. These two examples represent current ethical challenges in scientific research, but other examples of malpractice include the fabrication of data and wilful plagiarism. The temptations for fabricating or copying the perfect results may be great, but the potential damage that this can cause to both reputations and knowledge mean that the negatives vastly outweigh any wrongly perceived positives. As ethical scientists we have a responsibility to be vigilant of ourselves and others, and to be sure that we remain beyond reproach at all times.

Any research that you carry out should stand up to the ethical and integrity guidelines laid out by your research institute, especially if it involves the possible invasion of others' privacy. Most research institutions have such a policy, which sits within a hierarchy of legal and regulatory approaches to mitigate and respond to instances of malpractice. However, such ethical procedures are no longer the sole preserve of medical researchers and anthropologists, and must be taken extremely seriously whenever your research might have a direct influence on the lives of others; for example, by flying a drone near to people or a built-up area, or when using satellite imagery to record high-resolution imagery of privately-owned land.

As ethical scientists we must also act with integrity towards our fellow researchers. Avoiding plagiarism, explicitly seeking permission, and dispensing appropriate acknowledgement are essential ingredients for building a fertile research environment. If you are ever in doubt, then consider how you would feel if your own work had been abused in such a similarly anonymous manner. We owe it to each other, as scientists, to make sure that everyone is given a slice of cake that is proportional to what they have legitimately earned.

With this in mind, we have a duty to challenge and report instances of poor standards or academic integrity and malpractice. To ensure integrity as co-authors, we should challenge our own research teams if poor standards are identified, and work constructively to reach good solutions. As peer reviewers we have the same duty, and as members of the academic community, we should seek to publish a rebuttal of inaccurate work or conclusions. However, true malpractice may require a more formal and serious intervention. This can include reporting concerns such as plagiarism and fabrication to journals (which may lead to retractions for example), or reporting these same concerns to your institution, a funding body, a public or regulatory body, or in extremely rare circumstances, the police. When doing so, it may be useful to first discuss your approach to reporting malpractice with a trusted colleague or line manager, but this should not prevent you from acting on your instincts if you have an objective reason to seriously question a potential case of malpractice. For further reading on this very broad and important topic and its contemporary challenges, see the *Handbook of Academic Integrity* [22].

9.10 Promoting diversity

Section 9.4 discussed the benefits of building a diverse team, and why this is not just a box to be ticked but rather a useful and effective way to conduct scientific research. However, one of the biggest challenges to achieving this diversity is in overcoming our own prejudices.

Taking the Implicit Association Test (a version of which is available via Harvard University's Project Implicit® website [23]) is a necessary step to highlight any preference that you might have for certain social groups over others. While it may be uncomfortable to admit, very few of us are totally without prejudice, and research has shown that most people have an implicit and unconscious bias against members of disadvantaged groups [24].

Once you are aware of your own prejudices you can actively start to address them, and one of the ways in which this can be done is in helping to proactively promote and enable diversity. There are several steps that individual researchers can take in order to do this, most of which will depend on your personal circumstances, and often the extent to which you are 'privileged' within (and by) the scientific system. For example, as a male researcher you should turn down any invitations to join a 'manel' (i.e. male-only panel) and instead suggest several non-male colleagues to take your place. Strategies also include seeking and providing professional development to empower and equip members of your team to address their own unconscious biases [25].

If you are organising talks, seminars, and conferences, ensure that there is diverse representation. Crucially however, make sure that you are inviting these people to talk about their work and research, not just so that they can give an opinion on what it means to be a non-white-cis-heterosexual-male scientist. Providing a platform for a diverse range of scientists to talk about their research is a powerful way to help re-normalise science, i.e. to make diversity the norm rather than the exception.

9.11 Summary

This chapter has discussed a number of the additional skills and professional practices that are required in order to be a successful and an ethical scientist. Practical advice and activities have been provided that will help you to be more proactive in building a valuable and transferable skill set. Whatever direction your career is headed, you need to plan ahead, identifying areas in your expertise that require strengthening and actively seeking out ways to improve them. This can be done via training opportunities, professional development activities, or formal/informal mentoring. Take the time to build a robust and versatile CV, and use networking opportunities to find and develop contacts who will help you to maximise your true potential.

As scientists we are not just representatives for our research institutes and fields of research, but also for science in general. In conducting our research, we must approach all situations with high standards of integrity and consider the wider ethical implications of our work. Similarly, we have a responsibility to behave as ethical scientists, to acknowledge and address our own conscious and unconscious biases, and to proactively promote and support diversity in science.

9.12 Further study

The further study in this chapter is designed to help you think further about developing your essential research skills:
1. **Find a mentor.** From your five-year plan identify an area of expertise in which you require some assistance, be it either a technical skill or a more general one (e.g. grant writing, presenting, etc). Identify a colleague who has expertise in this skill, and ask them if they could provide you with some developmental advice. Offer to take them out for a coffee to discuss the matter at hand, and gradually start to sound them out for their opinion as you begin to develop your own expertise.
2. **Attend a course.** Your research institute will almost certainly offer a large number of Continuing Professional Development (CPD) opportunities, normally through its HR department. Again, use your five-year plan to identify areas in which you require training, and sign up for the appropriate courses. Where possible, it is always preferable to choose training opportunities that offer external accreditation, as these will be the most useful for future career opportunities.
3. **Get some experience.** If you have decided that a career in academic research is not for you, then identify opportunities to gain some relevant experience elsewhere. For example, if you want to go into teaching then volunteer at a local school. Similarly, if you are thinking about working in industry, then approach an appropriate company and arrange some knowledge exchange visits. In addition to bolstering your CV, gaining this experience will also help you to better decide if this is indeed the correct career path for you.
4. **Investigate open access.** Find out the approach to OA that is adopted by your research institute. If there is an OA group then ask if you can join, and if there isn't then think about setting one up yourself.
5. **Address you bias.** Take a version of the Implicit Association Test in order to better identify any unconscious biases. Then actively work to address these biases and to remove any prejudices in your approach to science.

9.13 Suggested reading

There are many web-based platforms that offer useful career planning tools for scientists and researchers, with the Vitae Researcher Careers website [26] offering a host of resources, which are useful if you want to either pursue an academic career, or use your skills elsewhere. The Institute of Physics also has some very valuable resources, including a hub for early career researchers [27].

If you want to find out more about open access, and the wider open science movement, then there are several recommended journal articles that present an accessible introduction to the subject, including discussions of required changes in research culture [28–30]. Finally, the journalist Angela Saini has written two very important books on addressing sexism (*Inferior* [31]) and racism (*Superior* [32]) in science, both of which are highly recommended and necessary reads.

References

[1] Taylor M, Martin B and Wilsdon J 2010 *The Scientific Century: Securing our Future Prosperity* (London: The Royal Society)
[2] Vitae Research Development Framework Planner https://rdfplanner.vitae.ac.uk/ (Accessed 16 October 2019)
[3] Goldstein D, Hahn C S, Hasher L, Wiprzycka U J and Zelazo P D 2007 Time of day, intellectual performance, and behavioral problems in morning versus evening type adolescents: is there a synchrony effect? *Personal. Indiv. Diff.* **42** 431–40
[4] Cirillo F 2018 *The Pomodoro Technique®: The Life-Changing Time-Management System* (London: Virgin Books)
[5] Bothello J and Roulet T J 2018 The imposter syndrome, or the mis-representation of self in academic life *J. Manage. Stud.* **56** 854–61
[6] Stevens F G, Plaut V C and Sanchez-Burks J 2008 Unlocking the benefits of diversity: all-inclusive multiculturalism and positive organizational change *J. Appl. Behav. Sci.* **44** 116–33
[7] Bear S, Rahman N and Post C 2010 The impact of board diversity and gender composition on corporate social responsibility and firm reputation *J. Bus. Ethics* **97** 207–21
[8] Roberge M-É and Dick R V 2010 Recognizing the benefits of diversity: when and how does diversity increase group performance? *Hum. Resour. Manage. Rev.* **20** 295–308
[9] Díaz-García C, González-Moreno A and Sáez-Martínez F J 2013 Gender diversity within R&D teams: its impact on radicalness of innovation *Innov.* **15** 149–60
[10] Cheruvelil K S, Soranno P A, Weathers K C, Hanson P C, Goring S J, Filstrup C T and Read E K 2014 Creating and maintaining high-performing collaborative research teams: the importance of diversity and interpersonal skills *Front. Ecol. Environ.* **12** 31–8
[11] Husebø S E, O'Regan S and Nestel D 2015 Reflective practice and its role in simulation *Clin. Sim. Nurs.* **11** 368–75
[12] Noveletsky-Rosenthal H T and Solomon K 2001 Reflections on the use of Johns' model of structured reflection in nurse-practitioner education *Inter. J. Hum. Car.* **5** 21–6
[13] Jay J K and Johnson K L 2002 Capturing complexity: a typology of reflective practice for teacher education *Teach. Teach. Educ.* **18** 73–85
[14] The Versatile PhD https://versatilephd.com/PhD-career-finder/ (Accessed 16 October 2019)
[15] Morais R, Stoy L S and Borrell-Damián L 2019 *Big Deals Survey Report 2019* (Brussels: European University Association asbl)

[16] King D W and Alvarado-Albertorio F M 2008 Pricing and other means of charging for scholarly journals: a literature review and commentary *Learn. Publ.* **21** 248–72
[17] Van Noorden R 2013 Open access: the true cost of science publishing *Nature News* **495** 426
[18] Project Gutenberg http://gutenberg.org/ (Accessed 16 October 2019)
[19] Science Europe *Seven Reasons to Care About Integrity in Research 2015* (Brussels)
[20] Sugarman J 2015 Ethics and germline gene editing *EMBO Reports* **16** 879–80
[21] Etzioni A and Etzioni O 2017 Incorporating ethics into artificial intelligence *J. Ethics* **21** 403–18
[22] Bretag T 2016 *Handbook of Academic Integrity* (Berlin: Springer)
[23] Project Implicit https://implicit.harvard.edu/implicit/ (Accessed 16 October 2019)
[24] Jolls C and Sunstein C R 2006 The law of implicit bias *Calif. L. Rev.* **94** 969
[25] Allen B J and Garg K 2016 Diversity matters in academic radiology: acknowledging and addressing unconscious bias *J. Am. Coll. Radiol.* **13** 1426–32
[26] Researcher Careers https://vitae.ac.uk/researcher-careers (Accessed 16 October 2019)
[27] Early-Career Physicists and Career Changers http://iop.org/careers/researcher--career-change/page_64175 (Accessed 16 October 2019)
[28] Masum H et al 2013 *Ten simple rules for cultivating open science and collaborative R&D* (Public Library of Science)
[29] Leonelli S, Spichtinger D and Prainsack B 2015 Sticks and carrots: encouraging open science at its source *Geo: Geog. Environ.* **2** 12–6
[30] Tennant J P, Waldner F, Jacques D C, Masuzzo P, Collister L B and Hartgerink C H J 2016 The academic, economic and societal impacts of open access: an evidence-based review *F1000Research* **5** 632
[31] Saini A 2017 *Inferior: How Science Got Women Wrong-and the New Research That's Rewriting the Story* (Boston: Beacon Press)
[32] Saini A 2019 *Superior: The Return of Race Science* (Boston: Beacon Press)

CPSIA information can be obtained
at www.ICGtesting.com
Printed in the USA
BVHW011103190720
583544BV00029B/19

9 780750 325189